ADVANCED FLOOD CONTROL SYSTEM WITH EMERGENCY POWER, ESCAPE ROUTES, AND CLEAN WATER:

AN EFFECTIVE LONG-TERM SOLUTION TO PREVENT MAJOR FLOODING IN LARGE MUNICIPAL AREAS

Abridged Version

by Mark Fennell
Oct 1, 2017

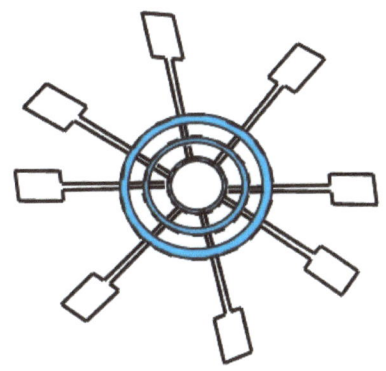

Table of Contents
for Abridged Version
of Advanced Flood Control Design

1. Executive Summary	5
2. Benefits of the Advanced Flood Control Design	7
3. For Full Version of Design and to Contact Designer	8
4. Basic Design of the Advanced Flood Control System	9
5. Basic Process of Flood Control	10
6. Effectiveness of the Flood Control System	11
7. Additional Solutions in the Flood Control Design	13
7a: Emergency Escape Tunnels	13
7b: Temporary Shelters	13
7c: Emergency Power Generators	14
7d: Clean Drinking Water	15
8. Combining Solutions in the Advanced System	17
Appendixes with Full Color Drawings and Descriptions	19
A1: General Overview Drawings	19
A2: Drainage Rings, Flood Chutes, Maintenance Tunnels	25
A3: End Point Systems	43
A4: Hydro-Power	61
A5: Filtration and Pumping Systems	67
A6: Maintenance and Power Lines	77
A7: Structural Boundaries of the End Point	91
About the Designer	97

Executive Summary
for Advanced Flood Control System

This Design is for an Advanced Flood Control System. This Design will effectively drain vast amounts of water from major metropolitan areas within a short time. This Design will also provide Emergency Power, Clean Drinking Water, and Additional Escape Routes.

In order to prevent major flooding in metropolitan areas, we need not just more pipes, but a completely new system. We need a system which will accomplish multiples goals, and solve multiple problems, all at the same time. The proposal in this document is such a system.

When installed properly, this design will manage all heavy rains effectively. Floods will never be a concern again.

Abridged Version

Note that this Document is an Abridged Version of the full design. This document provides the basic concepts, features, and benefits of the Advanced Design. However, there are numerous details which have been omitted.

Please read the Full Proposal for all details regarding this design. The Full Proposal is where you will learn why each design object is included, and how to build the entire system. Each detail is carefully laid out in that document, with descriptions and illustrations.

It is important for those building the system to fully understand how to do it, and why. Those details are provided in the Full Proposal, but are omitted in this Abridged Version.

For Further Information

For additional details on this Advanced Flood Management System, including the Full Design Proposal, please see the following:

1. Website: http://markfennellvisionary.com
2. Amazon: https://www.amazon.com/author/markfennellvisionary
3. You Tube: "All Things Energy"

Or contact the designer directly at: (512) 808-3446

Benefits of This Flood Control System

This Design is not just more pipes or entrance grates. Rather, it is an entire new system of flood management. This system provides not only efficient flood control, but so much more. Benefits of this Advanced Flood Control System include the following:

1. It will provide effective flood management for large areas, easily up to 150 miles in diameter.

2. It will effectively handle millions of gallons of water, in ways that quickly drain the streets.

3. Populations from thousands to millions will have their homes and businesses protected.

4. There will be emergency electrical power for thousands of residents throughout the region.

5. Drinking water will never be a problem, as there will be plenty of drinking water to serve all the region for several days.

6. Hospitals, Power Plants, Refineries, and Waste Sites will remain intact. These will never be damaged by water.

7. An emergency route is in place, with several entrance and exit points for the entire region.

8. The end points for the system are quite compact and arranged efficiently.

9. Most of the system should last 200 years or more without any repair.

10. Maintaining and replacing parts is easily done.

11. This system can be installed in most major cities, including: Houston, Dallas, Atlanta, Columbus, Cincinnati, Indianapolis, Kansas City, St. Louis, Prague, and many more.

Abridged Version

Note that this Document is an Abridged Version of the full design. This document provides the basic concepts, features, and benefits of the Advanced Design. However, there are numerous details which have been omitted.

Please read the Full Proposal for all details regarding this design. The Full Proposal is where you will learn why each design object is included, and how to build the entire system. Each detail is carefully laid out in that document, with descriptions and illustrations.

It is important for those building the system to fully understand how to do it, and why. Those details are provided in the Full Proposal, but are omitted in this Abridged Version.

For Further Information

For additional details on this Advanced Flood Management System, including the Full Design Proposal, please see the following:

1. Website: http://markfennellvisionary.com
2. Amazon: https://www.amazon.com/author/markfennellvisionary
3. You Tube: "All Things Energy"

or contact the designer directly at: (512) 808-3446

Basic Design of Advanced Flood Control System

The basic design of the system is a set of concentric drainage rings connected to flood chutes. It is essentially a wheel and spoke design.

Water will enter a drainage ring, flow to the nearest chute, then be swiftly carried away from the city. At the end of each chute will be a large underground container, which will store the water until the heavy rain has ceased.

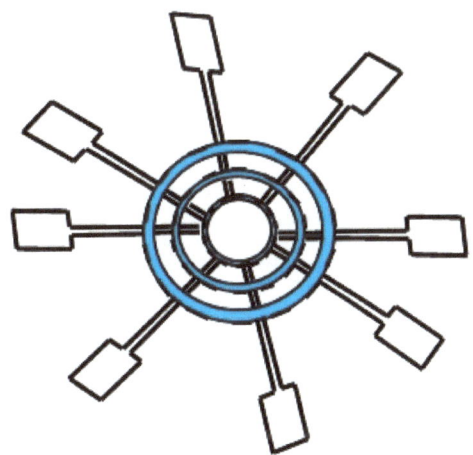

This Flood Control System primarily consists of:
- Concentric Drainage Rings
- Flood Control Chutes
- Storage Container Units

This System can also include:
- Escape Tunnel
- Emergency Power Generator
- Temporary Shelter
- Clean Drinking Water

Basic Process of Flood Control

The basic process of flood control in this system is as follows:

1. Water enters openings on the street
2. The water flows into one of the concentric Drainage Rings
3. Water then flows through the Drainage Ring to the nearest Flood Control Chute.
4. The Flood Control Chute then carries the water away from the city.
5. Outside the city, the water is dropped into a large Underground Storage Unit.

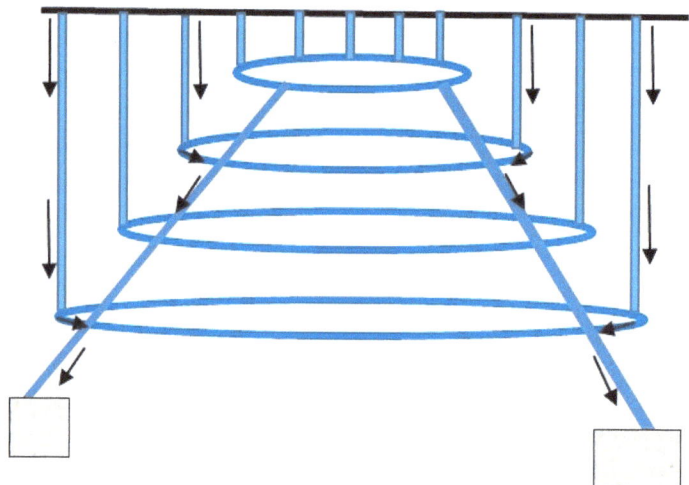

Effectiveness of the Flood Control System Design

The Geometric Design of this Flood Control System will be extremely effective in dissipating massive amounts of water, from all locations in the region.

1. The Geometric Design of the Ring and Chute System will exist in a wheel and spoke design. We begin with multiple concentric drainage rings. Extending across these rings – like spokes on a wheel - are the multiple flood control chutes. These chutes will carry the flood waters away from the city in multiple directions at the same time.

2. The Concentric Ring Design will allow all locations in the area to experience effective flood control at the same time.

    a) The Ring Pattern will allow multiple access points for water to enter the system

    b) The Concentric Rings will enable all locations, at different distances from the city center, to have their own Ring. This will enable each distance to have their own Drainage Ring.

    c) The Combination of the two geometries (Circular Drainage Rings, and Concentric Drainage Rings) will enable ALL REGIONS of the area to experience effective flood control. This means that every region in the territory will have access points for water to enter the drainage system. The result will be quick drainage from all streets in the metro area.

3. The Multiple Flood Control Chutes Allows for Efficient Flood Control

There will be multiple Flood Control Chutes, extending from the Drainage Rings, as spokes on a wheel. This design will allow water to be channeled from all regions most efficiently.

Because the Design is a wheel and spoke pattern, we can place Flood Control Chutes at multiple locations across the circles. This allows many chutes, which will enable water to be carried from multiple locations at the same time. With multiple flows of water, we can eliminate all water from the area very quickly.

Furthermore, because there are flood chutes at many angles, the water in the streets of all regions will quickly find access to the system.

4. Total Efficiency of the Flood Control Design

Taken together, the geometric design of this Flood Control System will provide quick drainage of flood waters in all locations of the area.

a) The circular drainage ring allows for multiple access points.

b) The concentric rings allows all distances to have access to system.

c) The multiple flood control chute allows water to be channeled away from multiple regions at the same time.

d) The multiple storage units makes storing the flood waters easier to manage.

Therefore, in total, this Design will manage the flood waters in an extremely efficient manner. With multiple access points, and multiple flood chutes, the flood water will be drained from all regions of the area very quickly.

In fact, the water will not become a flood at all; all water which comes to the streets will be drained away as soon as it arrives. There will *not be any flood*...just a continuous flow of water.

Additional Solutions in the Flood Control Design

In addition to the main flood control design above, we can easily incorporate other items into the system. There are many other solutions to practical concerns which we can incorporate into our design. These can be easily built at the same time as the Flood Control System above. These items include:

- Emergency Escape Tunnel
- Temporary Shelter
- Emergency Power Generator
- Clean Drinking Water

Here we will describe the basic concepts of these designs. We will also show how easy it is to incorporate these solutions into the main flood control system.

Emergency Escape Tunnels

An Emergency Escape Tunnel can be built in each community. These escape tunnels will be built parallel to the Flood Control Chutes. These Tunnels will also serve as the Maintenance Tunnel, allowing easy access for maintenance at any location.

The Escape Tunnels, as with the Flood Control Chutes, will lead out of the municipal area. There can be several entrance/exit points along the tunnel.

Temporary Shelters

Temporary shelters from major storms can also be built into the Flood Control System. These shelters can be placed near each of the entrance points.

In addition, the Escape Tunnel itself can serve as Temporary Shelter from the storms. People who live close to the access points can quickly go inside and stand within the tunnel itself, until the storms pass through. Using this method, the tunnel itself can serve as emergency shelter from major winds and rain for many people throughout the region.

Always remember that these are Temporary Shelters, only to protect people from the storm itself. It is not meant for long-term shelter.

Emergency Power Generators

Major storms will often destroy electrical power systems in the area. It is common for thousands of residents to be without power for days, or weeks, due to a major storm. This does not have to be the case.

Emergency Electrical Power Systems can be built around the city. These Emergency Electrical Power Systems will be part of the same Flood Control System described above.

The basic concept involves an underground hydropower system, using the rushing water from the chutes to provide the energy.

This Electrical Power System will be very simple to install. The hydropower turbines can be built directly on the chutes, just before the water falls into the storage container below.

The rushing water will flow from the chute, then through the turbine. This energy is transferred from turbine to the generator, where electrical power is created. Meanwhile, the water which has passed through the turbines falls directly into the storage container below.

Therefore, we can create electrical power quite easily, from every flood control chute in the region, from the natural energy of the rushing water. The result will be Emergency Electrical Power, created at the same time as the flood water are being contained.

Furthermore, this Electrical Power can be created from turbines on each of the flood chutes. Therefore, we will be creating Electrical Power throughout ALL regions of the city. The practical effect is that the entire city will have emergency electrical power.

Carrying the electrical power to the people will also be simple. The power created by these underground hydropower systems can be carried to the surface, using a series of underground cables. Note that these cables can be lined along the top of the Escape Tunnel; which makes a convenient and sturdy structure for the power lines as they carry power from generator to surface structures.

Therefore, these Emergency Power Systems are quite simple to install, and will be based on the same flood waters we are channeling from the city. These power systems will also be built along all flood chutes. Thus, again, the practical effect is that the entire city will have emergency electrical power.

Clean Drinking Water

One major problem that occurs after major storms is the lack of clean water for drinking. Millions of people are often left without adequate supply of drinking water. Yet with this Design the need for drinking water will never be a problem. There will always be enough drinking water for the entire community, throughout the entire recovery period.

The solution is a set of underground filtration systems. These filtration systems will transform some of the flood waters into clean water, then send that water back to the surface.

Essentially, we are using the heavy rain as drinking water. It is ironic that while the streets are flooded several feet deep, the city has no acceptable water for drinking. We will change that by adding a filtration system to the flood control system. This means that any amount of the rain coming to the region will not only be carried away, but also be used for drinking water.

These filtration systems will be built adjacent to the storage unit. We simply open a door, water enter the filtration unit. The unclean parts will be washed away to the adjacent dirt. The clean water will be pumped upwards to the surface.

Notice also that these filtration systems will be built at each water storage unit. There are several water storage units throughout the area (as part of the flood control system). Therefore, we will naturally have clean drinking water available to many locations in the region.

This Page
Intentionally Left Blank

Combining Solutions in Advanced Flood Control System

The beautiful aspect of this Flood Control Design, is that it is comprehensive. With one simple design, we can provide multiple solutions to significant problems all at the same time.

1. Flood Control
The main design allows efficient flood control, from all areas in the region. In fact, there will NOT be any floods, because the water is drained efficiently as soon as the rain arrives.

2. Emergency Hydropower
Electrical power will be created from each of the flood chutes. Water passes through the turbines, before falling into the storage container. Therefore we will have plenty of electrical power, for all areas, created by the very waters we are containing.

3. Clean Drinking Water
Clean drinking water will never be a problem again. Using a system of underground filtration units, we will be able to provide plenty of clean drinking water for all people of the city. The water supply will come from the very flood waters which we have channeled away, and kept in storage units around the region, which means there will be plenty of water to filter for the people.

4. Emergency Escape Routes, Shelters, and Power Lines
We can build a parallel structure to the flood chutes, which will serve as emergency escape routes. These tunnels can also serve as temporary shelters while the storm passes overhead. In addition, we can build underground rooms as temporary shelters.

These structures will also be useful supports for the Power Lines, which carry the electrical power from the hydropower systems to the surface above. The power lines will lay gently along the top, as the people walk past within.

Again, we are using one structure for multiple practical purposes.

The Comprehensive Design for the Advanced Flood Control System
Therefore, in total, this one Design for an Advanced Flood Control System, will easily provide multiple solutions to significant concerns, at the same time.

For Detailed Drawings, see Each Appendix

A1: General Overview and Main Concept Drawings of the Advanced Flood Control System

A2: Drainage Rings, Flood Chutes and Maintenance Tunnels

A3: End Point System Drawings

A4: Hydro-Power Drawings

A5: Filtration and Pumping Systems Drawings

A6: Maintenance, Power Lines, and Related

A7: Structural Boundaries and Support of the End Point

For Further Information

For additional details on this Advanced Flood Management System, including the Full Design Proposal with Construction Details, please see the following:

1. Website: http://markfennellvisionary.com
2. Amazon: https://www.amazon.com/author/markfennellvisionary
3. You Tube: "All Things Energy"

or contact the designer directly at: (512) 808-3446

Appendix 1:
General Overview and Main Concept Drawings for Advanced Flood Control System

Introduction

The Appendix pages have a series of drawings which illustrate the details of this Advanced Flood Control System. We begin with General Overview drawings. These pictures will illustrate the main concepts and the overall structure of the system.

Before presenting the illustrations, we will provide a brief review of the structure and operation of the Flood Control System.

Basic Design

The Basic Design of the Advanced Flood Control System is a series of concentric Drainage Rings, which are connected to a series of Flood Chutes, and lead to Flood Storage Containers.

The geometric layout is a wheel and spoke pattern, with multiple wheels, and multiple spokes. This layout is the most efficient for draining water from all areas of the region.

Note also that the depths are important. The Flood Chutes are sloped downward, for the water to flow naturally. Similarly, the Drainage Rings are buried at successive depths further from the city center; this is done in order to meet the Flood Chute at the exact depth at those coordinates.

Basic Flood Management Process

The water will easily flow from the streets to the storage container using the following process:
1. Rain falls to the street.
2. Water enters the nearest Entrance Grate.
3. Pipes carry water to the nearest Drainage Ring.
4. Water flows through Drainage Ring, to nearest Flood Chute.
5. The Flood Chute is the primary channel for the flood water.
6. Each Drainage Ring further from the city center will add more water to the Flood Chute
7. The flood water is eventually delivered to the Storage Container.
8. Multiple Flood Chutes, with multiple Flood Storage Containers, will allow all water from the region to be stored safely.

A.1.1
Basic Design of Concentric Drainage Rings and Multiple Flood Chutes

Schematic View, as Viewed from Above.

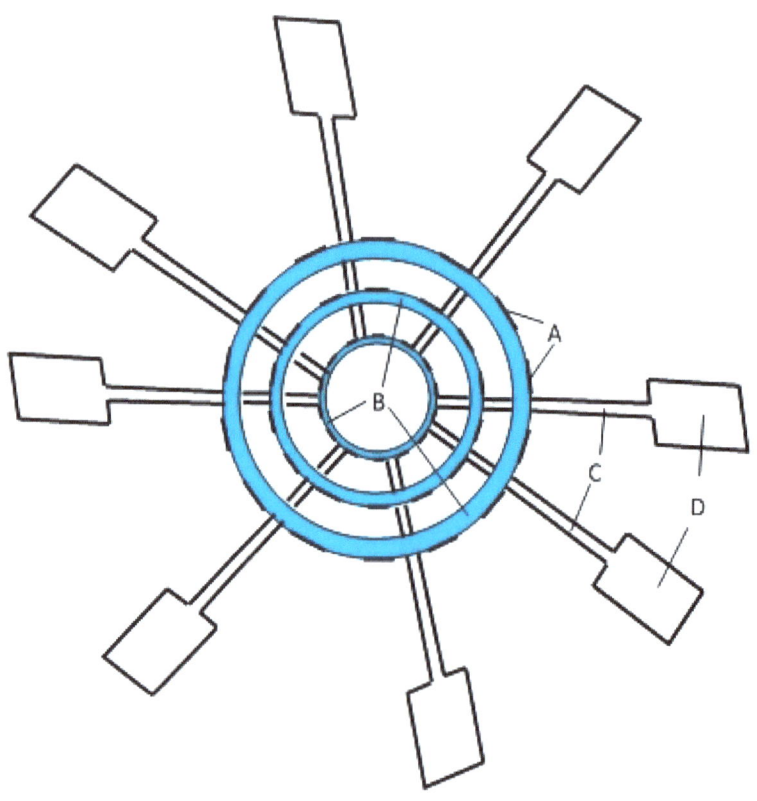

A = Entrance Grates

B = Drainage Rings

C = Flood Chutes

A.1.2
Side View of Flood Management System Underground.

Showing Increasing Depth of Concentric Drainage Rings, with Sloped Flood Chutes and Storage Containers

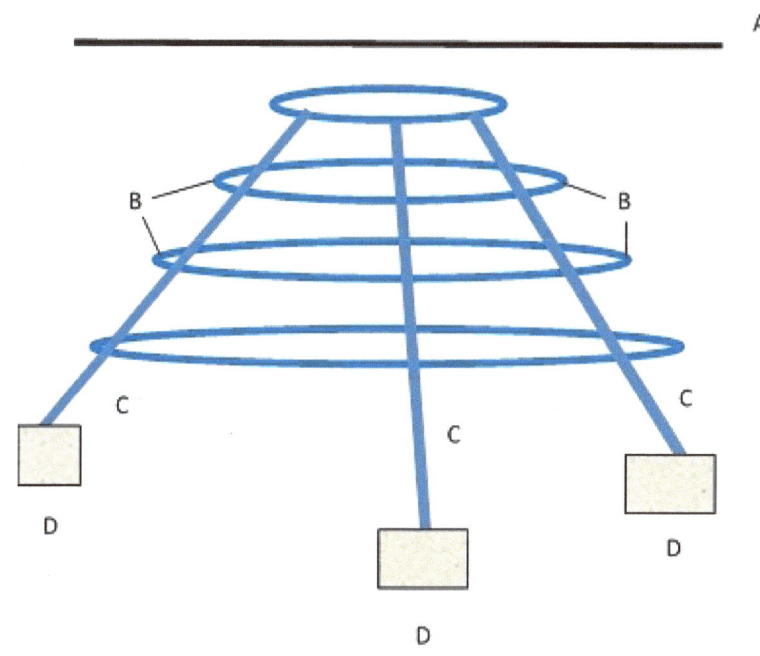

A = Ground Level

B = Drainage Rings

C = Flood Chutes

D = Flood Storage Containers

A.1.3
Side View of Flood Management System Underground

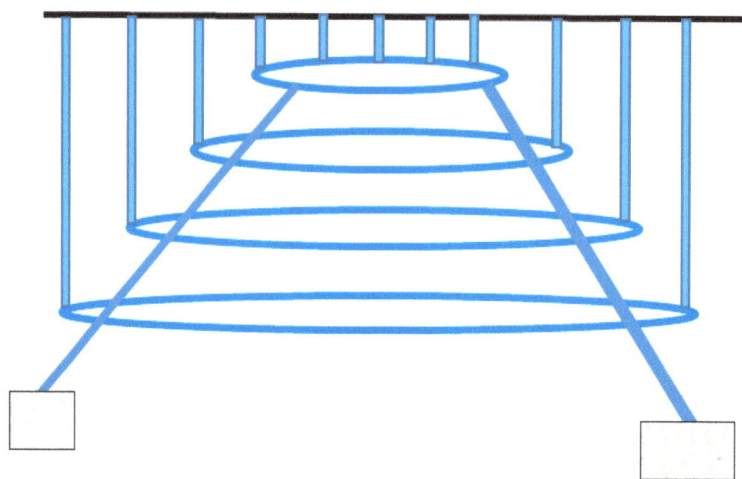

Showing Entrance Grates, First Pipes, Drainage Rings, Flood Chutes, and Flood Containers

A.1.4
Same as A.1.3, yet with arrows showing path of water

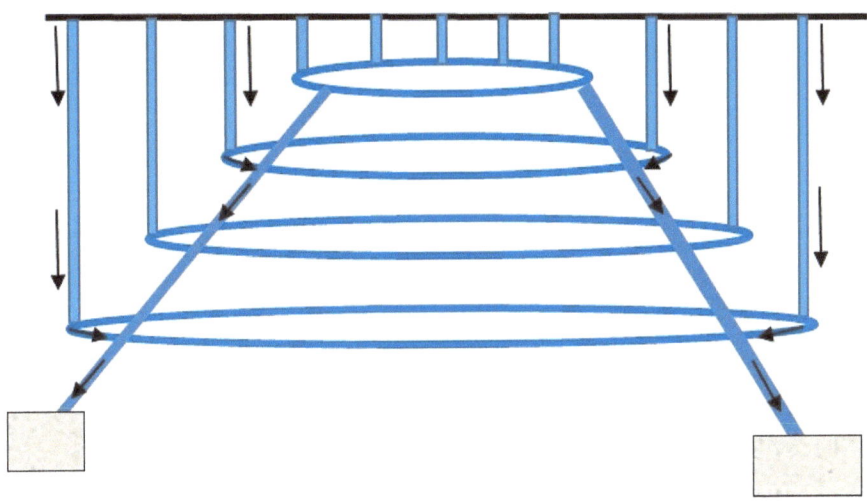

Note that only one line of Entrance Pipes is shown. In reality, these Entrance Pipes are installed as a full circle, directly above the drainage rings. Yet for simplicity, only one line of entrance pipes is shown here.

Additional Systems

Beyond the Flood Management System, there are other systems that can be easily built to co-exist at the same locations, and operated at the same times. These include: Hydropower, Filtration, Pumping, Maintenance, and Escape Routes.

These systems will be shown in other Appendix pages. However, we can place an overview picture here:

A.1.5 (also A.3.1)
Drawing of End Point System Altogether.
Top View, Looking Down

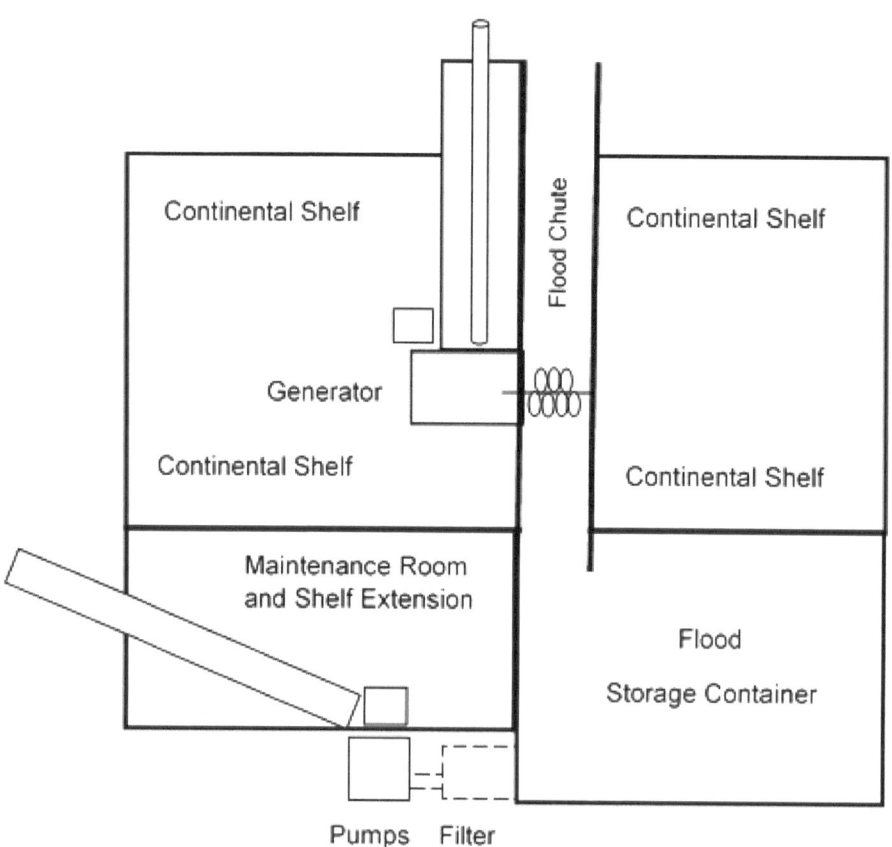

Appendix 2:
Drawings for Drainage Rings, Flood Chutes and Maintenance Tunnels

Introduction

The primary structures in the Flood Management System are: Drainage Rings, Flood Chutes, Storage Containers, and Maintenance Tunnels. Therefore the specific designs for these structures are important to build accurately. The following pages will illustrate the ideal designs for each of these structures.

Drainage Rings

Drainage Rings are large pipes, approximately 8 feet in diameter. These Drainage Rings are built in a concentric pattern around the city center. Multiple Entrance Grates and Connecting Pipes will bring the water from the streets to the Drainage Rings.

Because the Drainage Rings are circles, and because they are concentric, all locations in the region will receive efficient drainage of flood waters from the streets.

The large diameter will ensure that water flows quickly and easily. The flood waters will never be slowed, stopped, or reversed.

The Drainage Rings will also be built to varying depths, according to their distance from the city center. This will match the depth of the sloped flood chutes.

A.2.1
Drainage Rings built in Concentric Pattern, and to Varying Depths

A.2.2
Concentric Drainage Rings at Varying Depths, with Entrance Pipes, and Flood Chutes

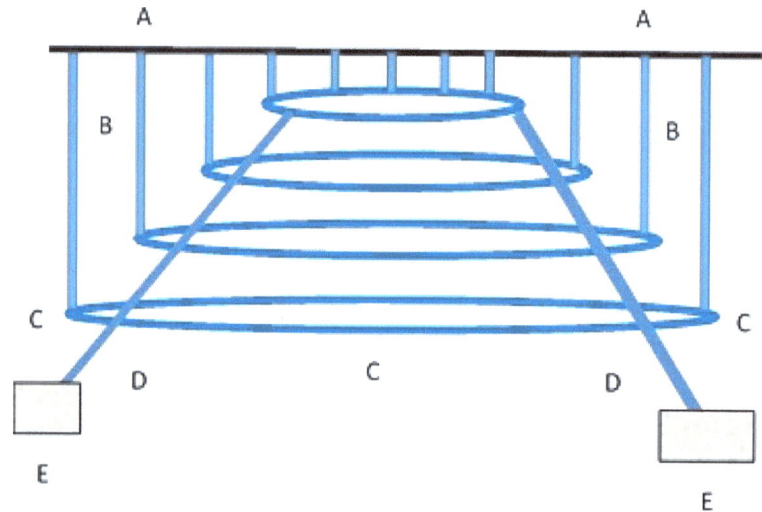

A = Entrance Grates

B = Entrance Pipes

C = Drainage Rings

D = Flood Chutes

E = Flood Storage Containers

Flood Chutes

The primary channels for moving the flood waters are the Flood Chutes. These Flood Chutes are wide concrete channels, which cross the Drainage Rings at several locations. Water from the Drainage Rings will flow into the nearest Flood Chute, then the Flood Chute will carry that water all the way to the Storage Container at the End Point.

The Flood Chutes are sloped at a gentle angle, to create a natural gravitational flow of the water.

The Flood Chutes are also built to increasing widths. As each Flood Chute passes a Drainage Ring, more water will be added, and therefore the Flood Chute must be widened to accommodate the total volume of water.

Several flood chutes will be built from various angles around the concentric Drainage Rings. The result will mean that all sections of Drainage Rings will find quick access to a nearby Flood Chute.

Each Flood Chute will ultimately carry the water to a Flood Storage Container outside of the city region.

A.2.3
Flow of Water in Flood Management System

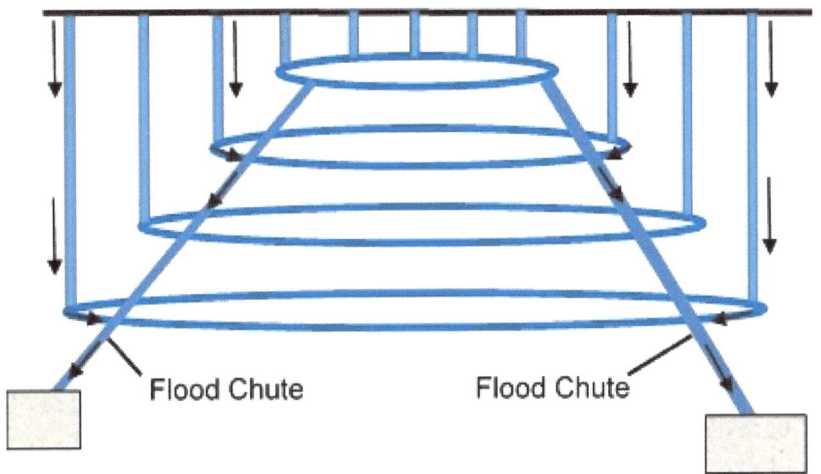

This Page Intentionally Left Blank

Ring-Chute Connectors

The Drainage Rings are connected to the Flood Chutes using the "Ring-Chute Connector Blocks". The connector blocks are square concrete blocks with parallel holes. The Drainage Pipes sit in the holes of the connector block, then into the holes in the walls of the Flood Chute.

A.2.4
Ring-Chute Connector Blocks

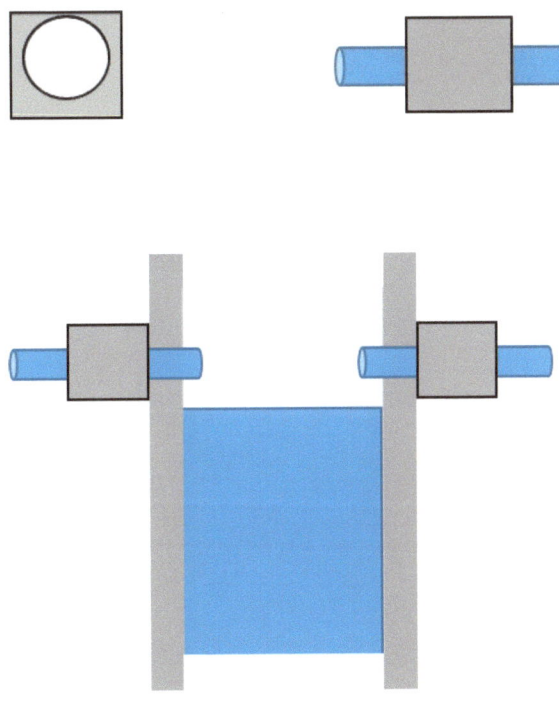

A.2.5
Flood Chute with Ring-Chute Connectors for Successive Drainage Rings

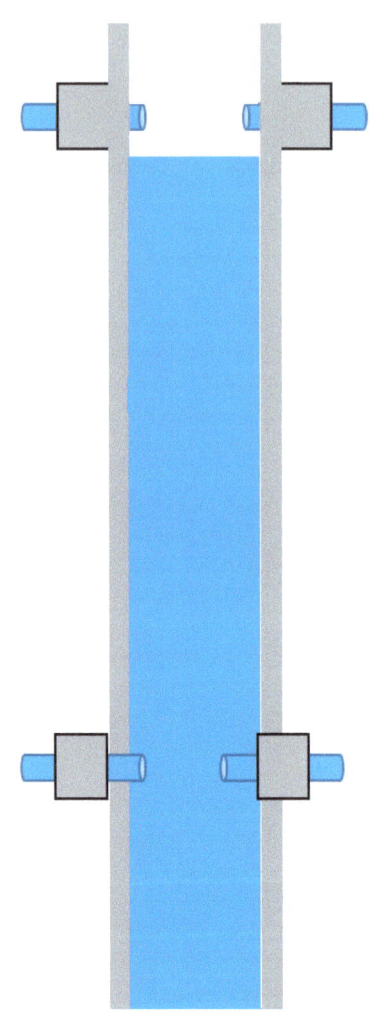

A.2.6
Increasing the Width of Flood Chutes after Each Successive Drainage Ring

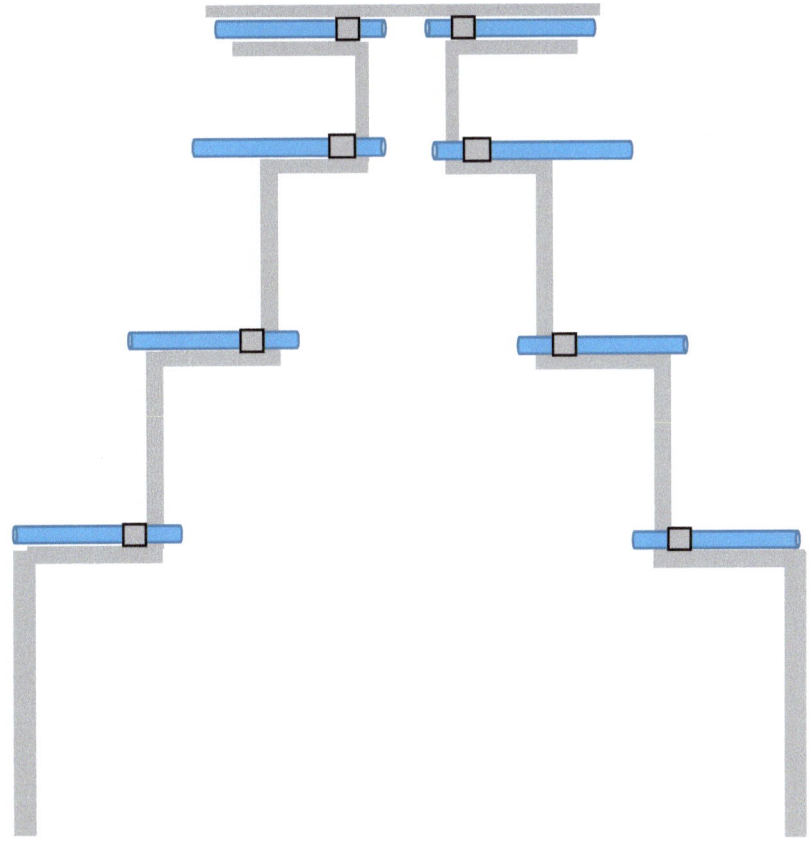

Flood Storage Containers

All flood waters will eventually be channeled into Flood Storage Containers. These Flood Storage Containers are large concrete vats, ideally lined with stainless steel.

Each Flood Chute leads directly to the top a corresponding Flood Storage Container, where the water falls into and becomes stored. These Flood Storage Containers are at the outer edges of the region covered, and are buried deep underground.

The number and sizes of these Flood Storage Containers will depend on the region covered, and the maximum rain volume in a flood period.

The water can be stored in these containers for a maximum of 1 month, by which time the water must be emptied (as agricultural water or drinking water). The Flood Container will then become available for the next rain storm.

A.2.7
Multiple Flood Chutes and their Storage Containers

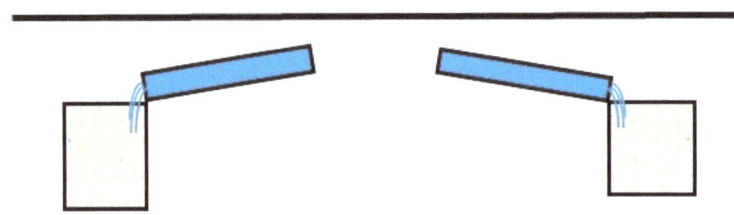

Surface Flood Chutes and Their Drainage Pipes

An additional option for Flood Chutes is to place them on the Surface. These operate as canals or artificial rivers, which then drain into the Drainage Pipes. These systems are intended to be in addition to the main flood control design, not as a replacement.

Note that there will be two types of Flood Chutes in this design. The first will be on the top, as Surface Flood Chutes. The second will be the original and main Flood Chutes, buried deep into the ground.

The Surface Chute will act essentially as a bathtub: the rain falls directly into the chute, then drainage holes will allow the water to flow into pipes below. The short pipes will carry water to the large Drainage Pipes, and to the main Flood Chutes beneath the earth.

This system is best for: flood plains, airports, prairies, swamps, and river beds. This system also works for rivers which may overflow.

See drawings A.2.8-A.2.10

A.2.8
Surface Flood Chute as it Crosses Several Drainage Rings Then to Nearest Underground Flood Chute

Note that this Diagram Spans Many Miles

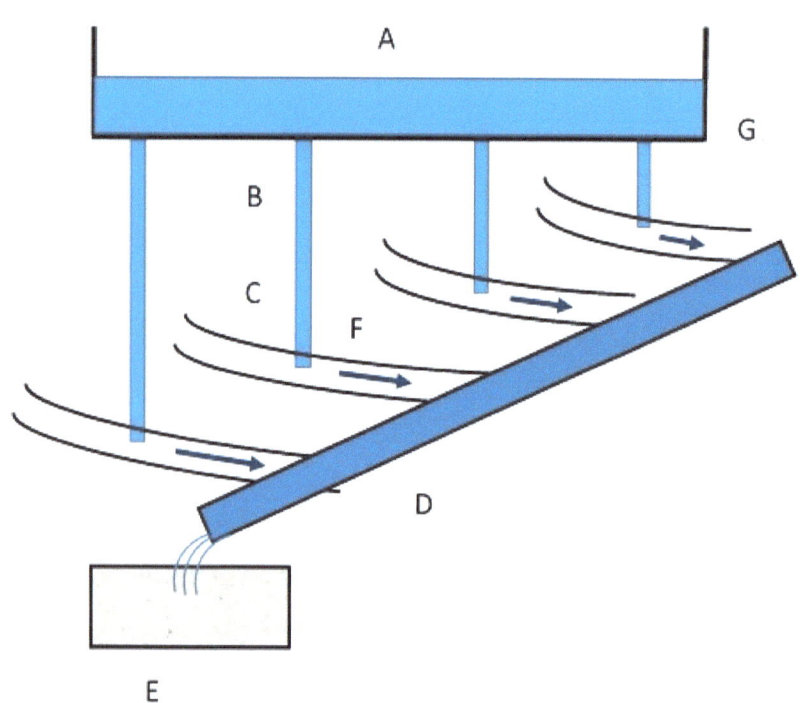

A = Surface Flood Chute
B = Drainage Connector Pipes
C = Drainage Rings
D = Underground Flood Chute
E = Flood Storage Container
F = Drainage Ring

Aerial Flood Chute with Drainage Pipes

Note that this entire system can also be raised in the air, with supporting posts or arches. This will become a traditional canal above the ground to carry rainwater to distant locations, while at the same time allowing for drainage pipes where needed.

Notice that the rainwater will be moved both laterally along the aerial chute, as well as downward though flood management pipes. This combination increases the efficiency of flood management significantly.

This modification is useful when crossing (and preventing flooding for) marshes, swamps, and farms. The supporting arches will step over the marsh or farmland, while the aerial flood chute and internal pipes will quickly move local rainwater to other locations. (See A.2.9)

A.2.9
Arch Support System for Supporting Aerial Flood Chute
Side View

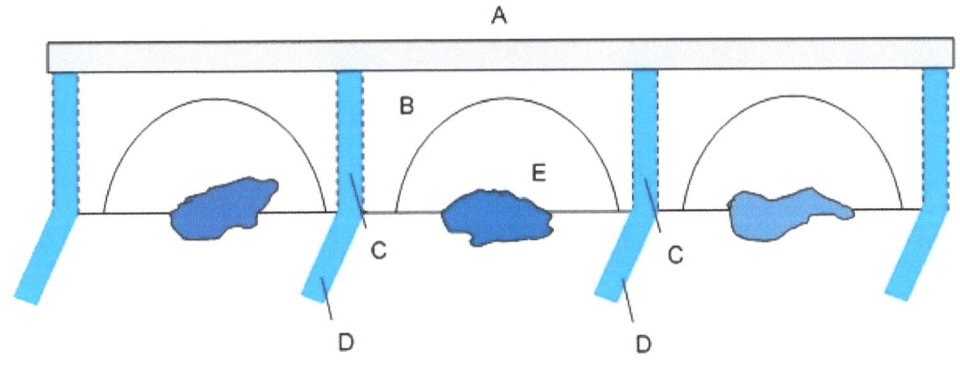

A = Aerial Flood Chute

B = Supporting Arches

C = Internal Drainage Pipes (Built inside the Arches)

D = Longer Distance Drainage Pipes

- Surface or Underground

- Eventually Leading to Flood Chutes or Local Storage

Surface Chutes Alternate Piping Design

An alternate design for the Surface Chutes is to place a tiered Drainage Pipe, which is in addition to the Drainage Ring.

In this design, there is a large pipe, similar to the Drainage Ring (but *not* the same object) placed directly below the Surface Chute. This Drainage Pipe is tiered directly below the chute, and follows the same path.

The water will enter this Drainage Pipe and flow underneath the Surface Chute. The water will travel many miles in this pipe, before finding the nearest Drainage Ring. More specifically, a connector pipe is placed from this Drainage Pipe to the Drainage Ring directly below.

The primary advantage to this method is to allow cross traveling paths of water, which will only intersect at certain points. The surface chute can travel in other directions (such as along a river, or across a marsh) than the traditional ring pattern. Yet intersect with the Rings when the geological features are acceptible.

Once the water has entered the traditional Drainage Rings, the water will flow to the nearest Underground Flood Chute (as in the basic design), and to the Flood Storage Container.

See diagram A.2.10.

A.2.10
Surface Flood Chute Tiered on Parallel Drainage Pipe Traveling for Distance, then into Traditional Drainage Ring

Side View of Surface Chute Segment near Drainage Ring Pipe

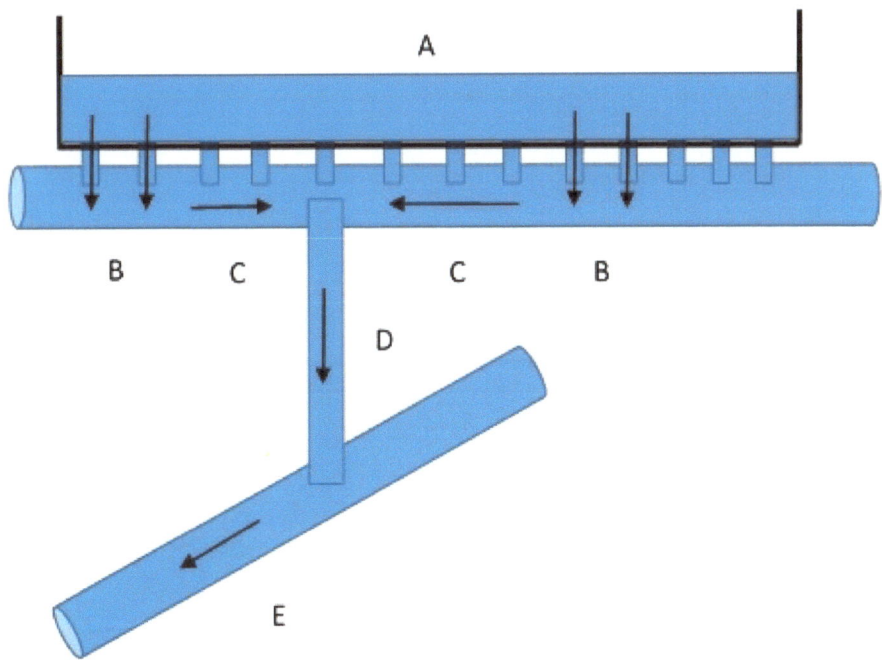

A = Surface Flood Chute

B = Entrance Pipes
   from Surface Chute to Horizontal Drainage Pipe

C = Tiered Parallel Drainage Pipe

D = Connection Pipe
   from Horizontal Drainage Pipe to Closest Drainage Ring

E = Traditional Drainage Ring

This Page Intentionally Left Blank

Maintenance Tunnels and Escape Routes

Maintenance Tunnels are an important part of the Flood Management System. In general, the Flood Chutes should require very little maintenance. However, some maintenance will be required, and this can be easily done using the Maintenance Tunnels. These Maintenance Tunnels will also serve as Emergency Escape Routes.

The Maintenance Tunnels will run parallel to the Flood Chutes, and in fact will share one wall. They will be both be installed at the same gentle slope.

Maintenance Tunnels will share one wall with the Flood Chutes, however the Flood Chutes will generally be taller (for all the water which must be contained). The bottoms of the Maintenance Tunnel and Flood Chutes will be identical, yet the heights will differ.

The Ring-Chute Connectors and the inserted Drainage Rings will pass through the Maintenance Tunnel on their way to the Flood Chute. In order to accommodate this obstacle for the workers, a gradual hump is created over the top of the Ring-Chute Connector.

Maintenance Doors and Lights are placed throughout the Tunnel, powered by an individual line from the generator at the End Point.

In addition, there is a Main Power Line which is laid on the roof of the Maintenance Tunnel, on its way to the reach the people. The placement of the Power Cable on the roof of the Maintenance Tunnel is a convenient way to gradually bring the power cable to the city center.

See figures A.2.10 and A.2.11.

A.2.10
Maintenance Tunnel / Escape Route

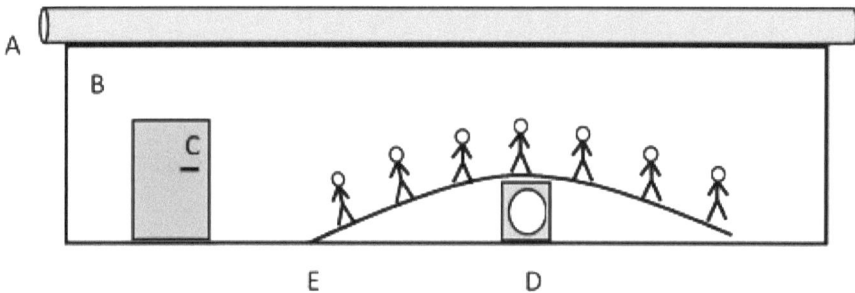

A = Main Power Line

B = Maintenance Tunnel

C = Maintenance Door to Flood Chute

D = Ring-Chute Connector

E = Ramp over Ring-Chute Connector

A.2.11
Maintenance Tunnel / Escape Route Details Shown Next to Flood Chute

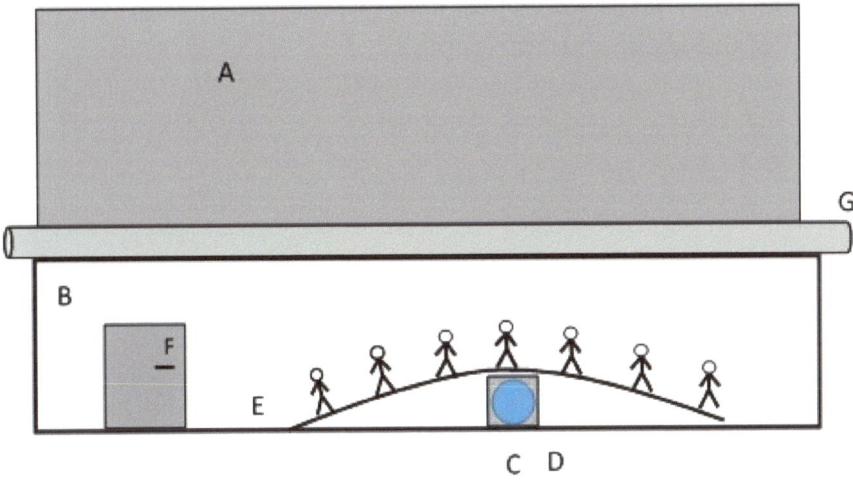

A = Flood Chute
B = Maintenance Tunnel
C = Drainage Ring
D = Ring-Chute Connector
E = Ramp over Ring-Chute Connector
F = Maintenance Door to Flood Chute

    The water flows through the Drainage Ring (C), which is set inside the Ring-Chute Connector (D). The water then flows into the Flood Chute (A). The Flood Chute will then take the water all the way down to the Storage Container (not shown).

    Adjacent to the Flood Chute (A) is the Maintenance Tunnel (B). Notice that the Flood Chute and Maintenance Tunnel shares one wall. Notice also that the Flood Chute is usually much taller than the Maintenance Tunnel.

People can walk over the Drainage Ring (C) inside the Maintenance Tunnel by using a gradual ramp (E) over the Ring-Chute Connector (D). Maintenance Workers access the Flood Chute via Maintenance Doors (F).

On top of the Maintenance Tunnel is the Main Power Line. Just before the water drops into the storage container the water passes through a hydropower system. The power created will be brought up in a conduit (G) which is gently laid on top of the Maintenance Tunnel.

Appendix 3:
End Point Underground Systems

Note that this is an Abridged Version of Appendix 3

Introduction

The "End Point" of this Flood Control System is where many important things happen. Therefore, deep in the ground, below a meadow in a rural area, the following activities occur:

- Flood Waters are dropped into the Storage Container
- Hydropower is created.
- Water is filtered into drinking water.
- Clean water is pumped to the surface.

Additionally:
- An elevator brings equipment down, and people up.
- Maintenance Tunnel has its end point.
- Power Line are placed to distribute power as needed.

Therefore, much goes on in this one location. And yet, the arrangement of all objects is designed for maximum efficiency, ideal simplicity, and multiple uses where possible.

The drawings on the following pages will illustrate how all of the objects can be arranged, sized, and built together for this maximum effectiveness.

About the Drawings

These drawings will show the arrangements of all objects that exist at this one location. Remember that underneath the Final Meadow will be the End Point for the flow of flood waters. It is here that the Storage Container will store the flood waters…and so much more. Also remember that these End Point Systems will be placed at the end of every Flood Chute. The drawings on the following pages will illustrate the ideal arrangement and sizing of all objects at this location.

Important Terms Used in these Drawings

Because there are so many important components in the End Point Underground System, we must clarify the terms before presenting the drawings:

1. "End Point Underground System"

This term refers to the entire set of components and subsystems at the End Point. This includes:
- Flood Management System
- Hydropower System
- Filtering System
- Pumping System
- Maintenance System
- Power Line System

2. "Continental Shelf"

The "Continental Shelf" is the large flat surface on which generators and other equipment sit, far above the storage container below.

The End Point Underground System is constructed similar to the Continental Shelf in the oceans. In the ocean, there is a long surface (the continental shelf), then a sharp drop into the trench.

Similarly, in our End Point, we have a long flat surface, before the drop into the storage container. On this large surface sits many things, such as the Generator, the Maintenance Tunnel, the Elevator, and a few Pumps. This large surface, before the drop into the storage container, we will refer to as the "Continental Shelf" of the End Point System.

3. "Main Level"
The "Main Level" of the End Point System is the level where the majority of objects are placed. It is also the top level of several structures.

The "Main Level includes: The Continental Shelf; the Surface Extension (top of Maintenance Room), and Top of the Flood Storage Container.

Contrasting levels include: Absolute Ground, and On the Main Level. The next level below the Main Level is the Absolute Ground (see below), and includes the filter. Objects which sit "On the Main Level" include the Generator, Maintenance Tunnel, and the Elevator.

4. "Absolute Bottom"
The "Absolute Bottom" is the lowest level of our End Point. There will be nothing built deeper than this level.

Continuing on the Continental Shelf vs Trench comparison, we have an absolute bottom for our End Point, similar to the bottom of a trench in the ocean.

5. "Storage Container"
The "Storage Container" is the primary Flood Management Container. All the waters which flow through the Drainage Rings the Flood Chutes will ultimately be dropped into the Storage Container.

Note that using the "Continental Shelf" analogy, the Storage Container is our "trench".

6. "Long-Term Storage"
The "Long-Term Storage" is where we store clean drinking water, for long periods of time. These storage containers are on the surface, or just below.

Note the differences between the Storage Container and Long-Term Storage.

- a) Long-Term Storage contains clean filtered water; Storage Container has water from the street, not yet filtered.

- b) Long-Term Storage is on the surface or just below; Storage Container is deep in the ground.

7. "Surface Extension" and "Maintenance Room"

The "Surface Extension" has the same height as the Continental Shelf, but of completely different design. Whereas the Continental Shelf is one giant slab of concrete (like a large floor), the "Surface Extension" is in fact a full box shape.

This "Surface Extension" has walls of concrete, reaching from the Absolute Bottom to the same height as the Continental Shelf.

The surface of the "Surface Extension" exists primarily to support the first Archimedes Screw; this is an ideal height to begin the conveyance. The surface being equal to the Continental Shelf will also provide maintenance simplicities.

The interior is hollow. This is a multi-purpose use of the Shelf Extension, which acts additional storage. This can be used as a Maintenance Room, or as Additional Water Storage for flood waters if needed.

8. "Maintenance Room" or "Storage Room"

The "Maintenance Room" is a large multi-purpose room, not just for the interior but for the top surface as well.

As described above, this "Maintenance Room" is a concrete box. The top of the Maintenance Room is also known as the "Surface Extension", where important objects can sit. The structure being a concrete box will ensure the stability of all objects on the Surface Extension, including the Archimedes Screw.

At the same time, the interior of this concrete box is hollow. This will allow for additional storage. This storage can be used for Maintenance Equipment, including replacement parts for filters, pumps, and turbines.

This storage area may also be used for additional flood water if necessary. The main "Flood Storage Container" is of course designed to be large enough to handle all the flood water. However if the water is greater than the capacity of the storage container, then this Maintenance Room may be opened up to allow additional water to be contained temporarily.

9. "Pump" or "Standard Pump"

This refers to any mechanical pump which can send water upwards, typically at large volumes and significant heights. Numerous options exist.

Important to note that these differ from "Archimedes Screws", which are the other main mechanisms in this design for carrying water to the surface.

10. "Archimedes Screw"

The "Archimedes Screw" is a well-known device for conveying water. I Turning the screw will convey the water gradually upward. This device is used in many modern pumping and power systems today.

Codes: Dashed Lines vs Solid Lines

In order to better understand the arrangement of these objects, both horizontally and vertically, these drawings use a combination of solid lines and dashed lines.

A) <u>For Top Views (looking down)</u>:
   1. Solid Lines indicate object is at Highest Level.
   2. Dashed Lines indicate roof of object is Deeper.

B) <u>For Side Views</u>:
   1. Solid Lines indicate object is in Front.
   2. Dashed Lines indicate object is Behind Another.

<u>Example: Top View</u>

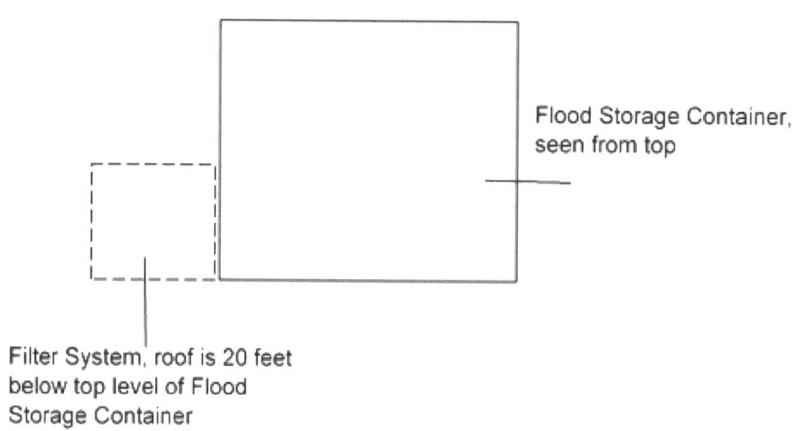

Flood Storage Container, seen from top

Filter System, roof is 20 feet below top level of Flood Storage Container

Example: Side View

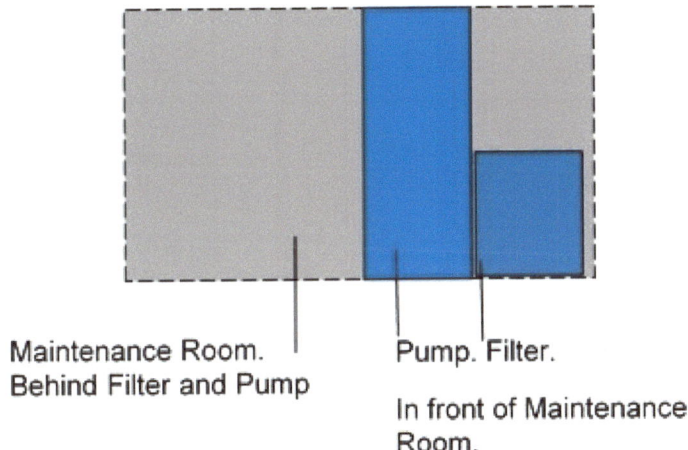

Maintenance Room. Behind Filter and Pump

Pump. Filter. In front of Maintenance Room.

A.3.1
Drawing of End Point System Altogether
Top View, Looking Down

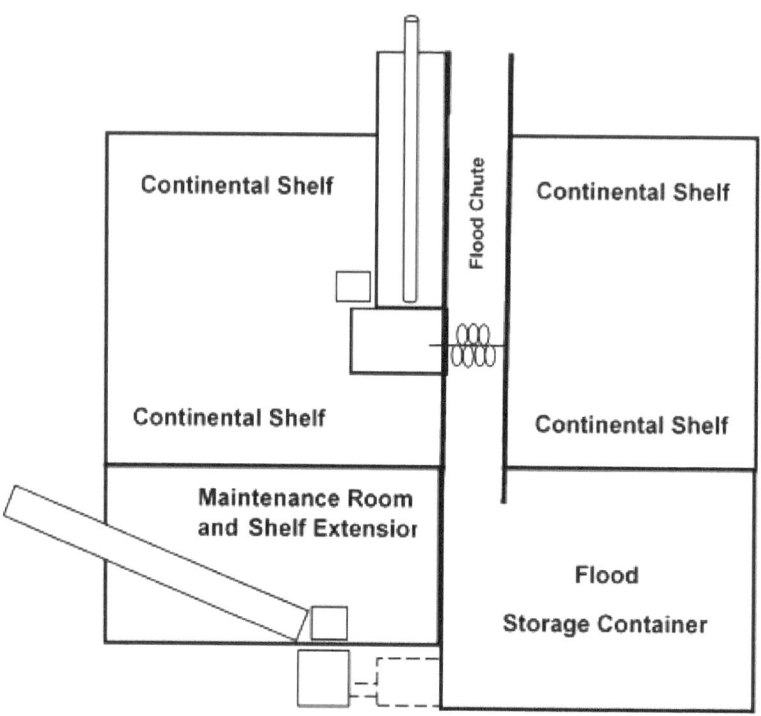

In this drawing you will notice the following:

1. The Main Level consists of: the Continental Shelf, the roof of the Maintenance Room, and the top of the Flood Storage Container.

2. The Filter may appear to be adjacent to the Storage Container. However the filter is actually placed much deeper than the Storage Container. Horizontally, they are adjacent, but vertically the filter is much deeper.

3. On top of the Continental Shelf are placed: Flood Chute; Maintenance Tunnel; Generator; and Elevator. Further, on top of the Maintenance Tunnel is the Main Power Line that leads to the surface.

4. On top of the Surface Extension/Maintenance Room is the Archimedes Screw, and the Gear Box to operate it.

Additional details will be shown in drawings below

A.3.2
Objects Arranged On Top of the Continental Shelf.
Top View, Looking Down

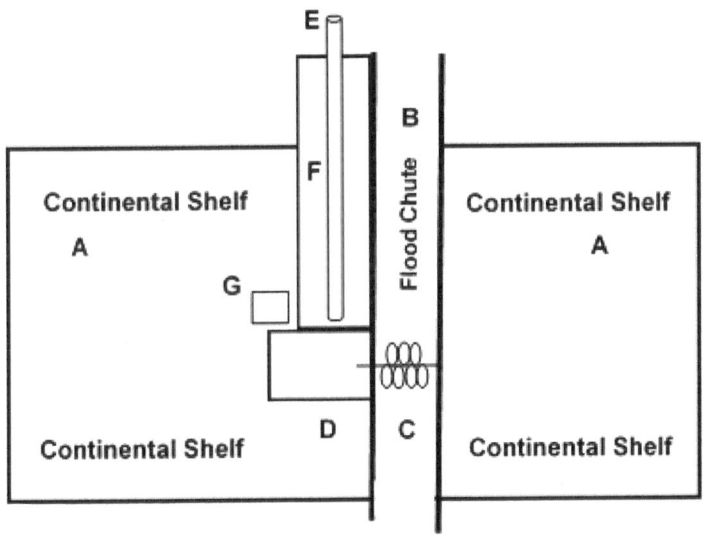

A = Continental Shelf
B = Flood Chute
C = Turbine
D = Generator Room
E = Main Power Line
F = Maintenance Tunnel
G = Elevator

*Note that the Flood Chute and the Maintenance Tunnel are slightly sloped as they come from the back of the Continental Shelf to the front. However, the Continental Shelf remains flat.

A.3.3
Filtration and Pumping Equipment, Arranged Together. Top View, Looking Down

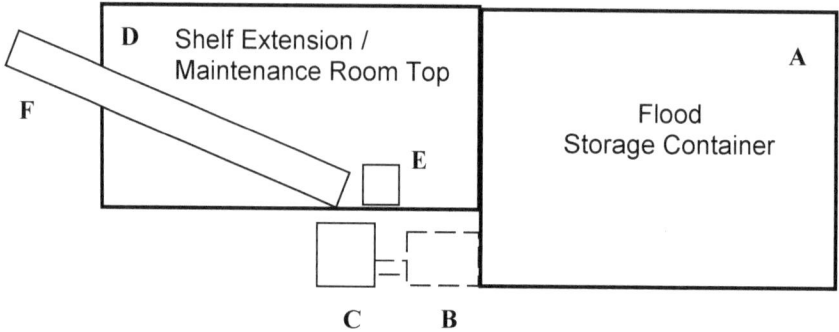

A = Flood Storage Container

B = Filter

C = Primary Pump

D = Maintenance Room Top / Shelf Extension

E = Gear Box for Archimedes Screw

F = Archimedes Screw

A.3.4
Side View of Filter, Pump, Flood Container, and Maintenance Storage

**Note this is viewed from standing at "Absolute Bottom" level. This is the filter and primary pump system you will see from that perspective.

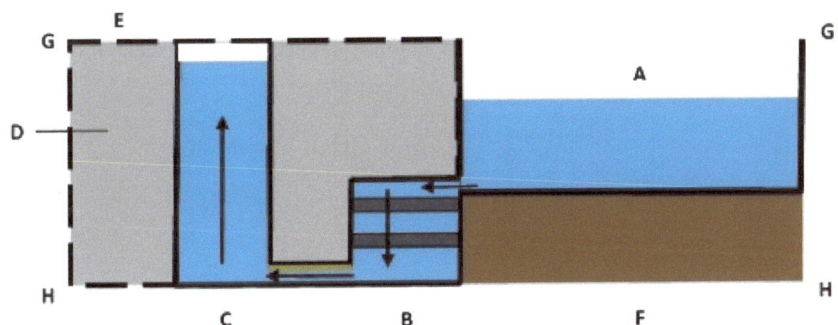

A = Flood Storage Container

B = Filter

C = Primary Pump

D = Maintenance Storage Room (Behind the Filter and Pump)

E = Surface Extension (Top of Storage Room)

F = Dirt and Rocks

    (Flood Storage is on dirt, a few feet above Absolute Bottom. This is for filter to work.)

G = Main Level

H = Absolute Bottom

A.3.5
Generator Room Option A

Omitted in Abridged Version

A.3.6
Generator and Power Line System for Option A

Omitted in Abridged Version

A.3.7
Generator Room Option B:
Two Turbines, Two Generators, 3 Phases x 2.
Top View

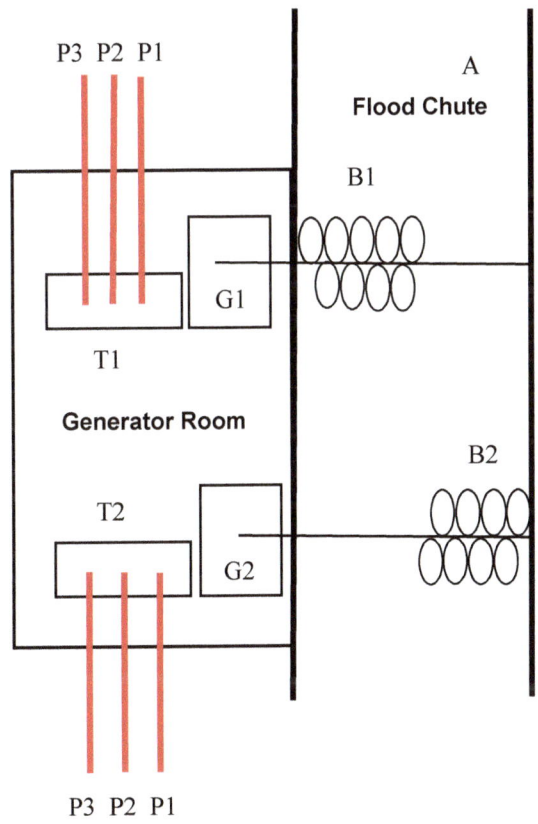

A = Flood Chute
B1 = Turbine 1   B2 = Turbine 2
G1 = Generator 1   G2 = Generator 2
T1 = Transformer Bank 1   T2 = Transformer Bank 2

P1, P2, P3 = Phase 1, 2, 3 from each Generator

(1) All 3 phases of Power from Generator 1 will go up conduit, providing power for the public use.

(2) Generator 2 provides power for internal operations:
    P1 = for Primary Pump and Deeper Operations
    P2 = for Lights, Doors, Elevator
    P3 = for Archimedes Conveyances and Other Pumps

A.3.8
Generator and Power Line System
Two Turbines, Two Generators, 3 Phases x 2
Top View, Looking Down from Above

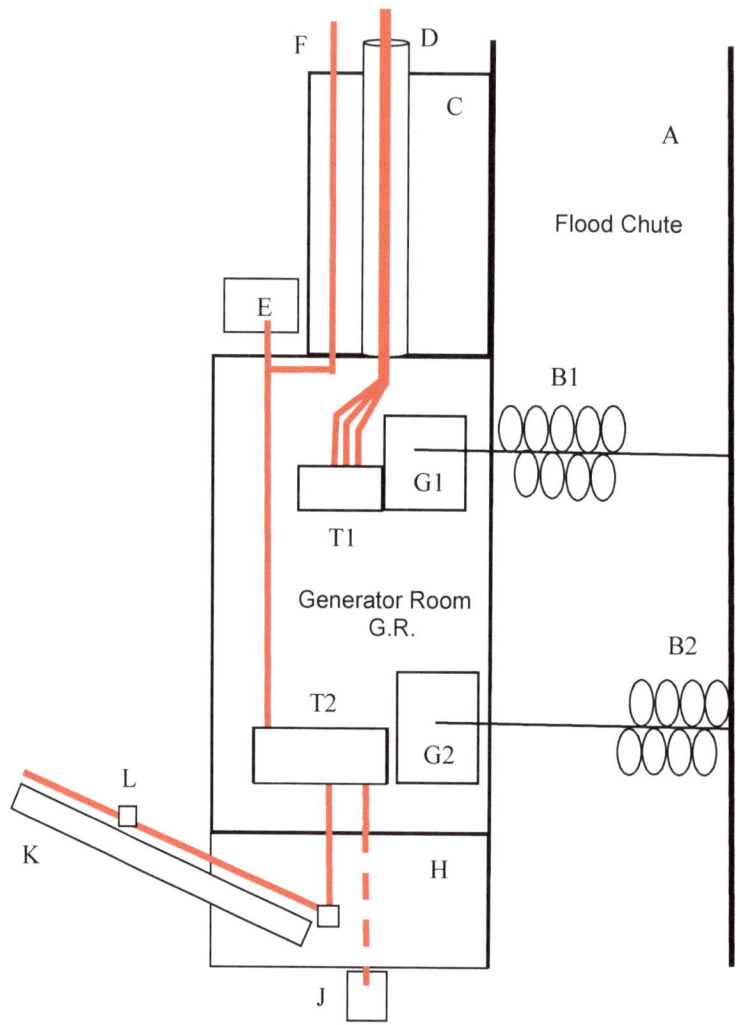

For meanings of each letter in this Diagram, see next page.

The Letter Meanings for Drawing A.3.8

A = <u>Flood Chute</u>
B1 = Turbine 1   B2 = Turbine 2

G.R. = <u>Generator Room</u>
G1 = Generator 1   G2 = Generator 2
T1 = Transformer 1   T1 = Transformer 2

C = <u>Maintenance Tunnel</u>
D = Conduit on roof, with Main Power Line
E = Elevator
F = Power Line for lights and doors in Maintenance Tunnel

H = <u>Shelf Extension / Top of Storage Room</u>
J = Primary Pump (at Absolute Ground, in front of Room)
K = Archimedes Screws for Water Conveyance
L = Gear Box for each Archimedes Screw

---

(1) All 3 phases of Power from Generator 1 will go up conduit, providing power for the public use.

(2) Generator 2 provides power for internal operations:
    P1 = for Primary Pump and Deeper Operations
    P2 = for Lights, Doors, Elevator
    P3 = for Archimedes Conveyances and Other Pumps

A.3.9
Pulley Platform to Absolute Bottom and the Access Doors to Storage Room.

Side View, *Outside* of Storage Container

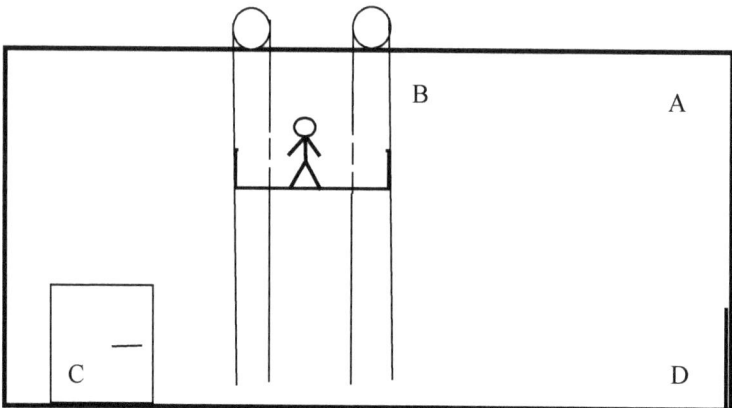

A = Maintenance Storage Room

B = Pulley System for Delivering Men and Equipment
(from Main Level to Absolute Ground and back up)

C = Door to Maintenance Storage Room

D = Door between Maintenance Room and Flood Storage Container (accessed inside Storage Room)

A.3.10
Access Doors and Pulley Systems
for Pump, Filter, Flood Container.
Viewed from Side

*Remember that Pulley Systems are on the *outside* of the Maintenance Storage Room. Also remember that the Filter and the Primary Pump will sit in front of the Storage Room.

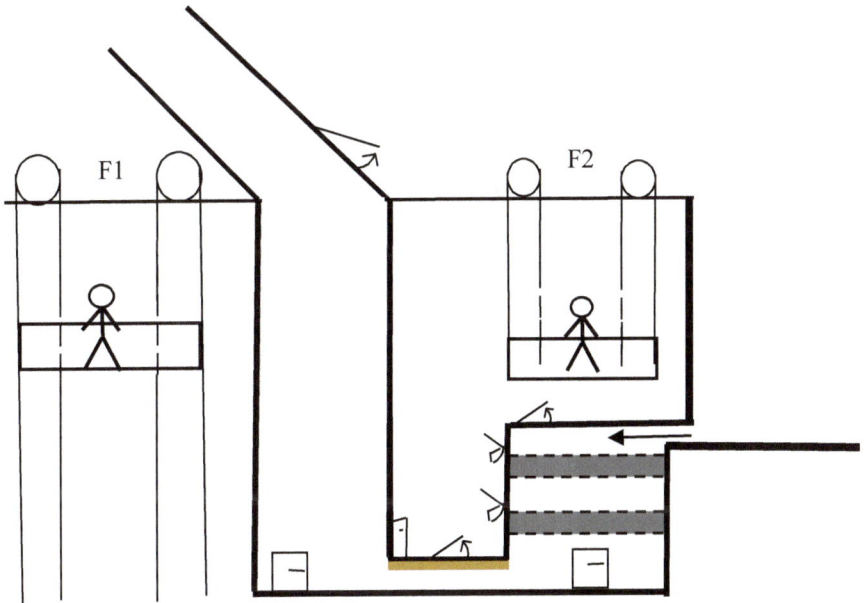

A = Flood Storage Container
B = Maintenance Storage Room
C = Filter System
D = Primary Pump
E = Archimedes Screw Conveyance

F1 = Pulley System to Absolute Ground
F2 = Pulley System to Top of Filter

G = Main Door to Pump Structure
H = Main Door to Filter Structure
J = Doors to Filters and Filter System
K = Side Door to Primary Pump
L = Maintenance Door to Archimedes Screw

This Page Intentionally Left Blank

Appendix 4:
Hydropower Drawings

Introduction

Using Hydropower in this Advanced Flood Control System is relatively simple, and yet very effective.

We use the flow of storm water through the Flood Chutes as energy for the hydropower system. There is enormous amount of energy in the volumes of water sent through these flood chutes during the storm. We simply install turbines and generators along the flood chutes, then the flowing water will generate the electrical power, which we can then deliver up to the surface to the people.

Generator Rooms and Power Lines

Generator Rooms can be placed either to the side of the flood chutes, or below the flood chutes. The preferred placement is to install generators to the side of the chutes. The Generator Rooms will have generators, transformers, and resulting power lines.

Every flood chute should have one generator, and preferably two. The first generator will provide the main power for the public. The second generator will provide local power, for pumping systems and maintenance rooms.

The main power line will be led up to the public on the top of the maintenance tunnel. The other power line will be divided to power each component of the pumps, doors, and lights underground.

Turbines

Turbines can be installed to rotate either horizontally or vertically. The choice should depend on where the generator rooms are located.

Note that turbines are always manufactured for the specific site, always custom made. Therefore designing the rotation of the turbine, along with the proper blade styles, will be easy to accomplish.

The turbines are best placed inside the flood chute. They can be placed anywhere along the chute, however the preferred location is approximately 100 feet before the drop off into the storage container.

A.4.1
Generator Room version A:
Generator on Left Side of Flood Chute

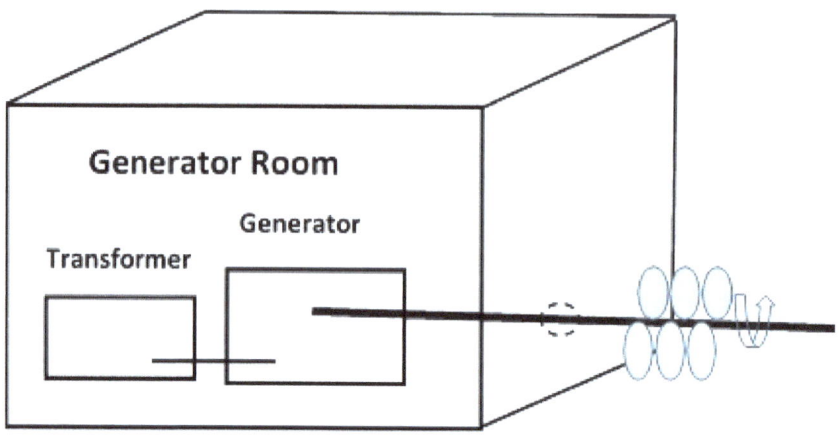

A.4.2
(Omitted in Abridged Version)

A.4.3
Generator Room B: Below Flood Chute

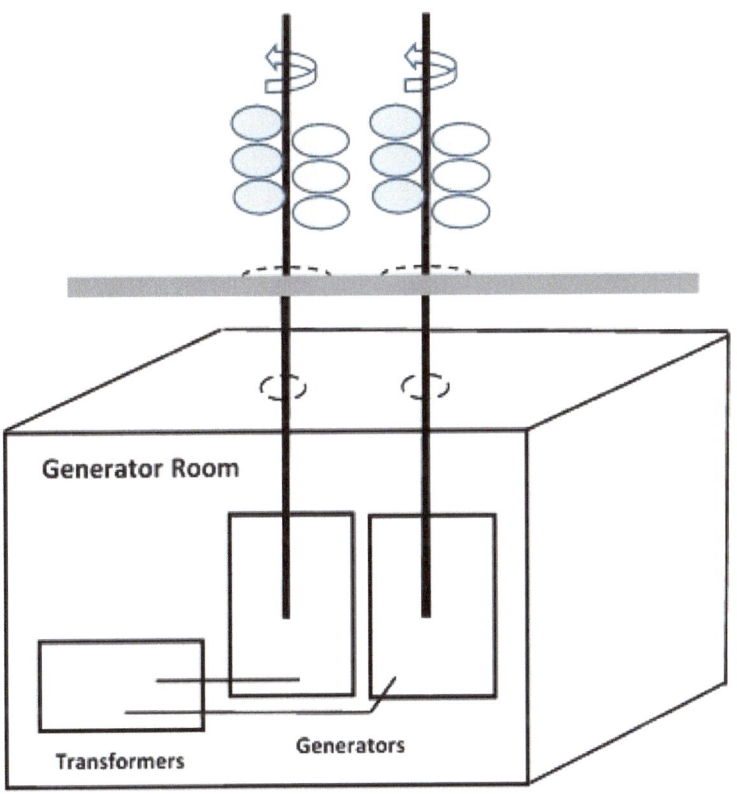

A.4.4
Hydropower System version B: Generator Below Flood Chute

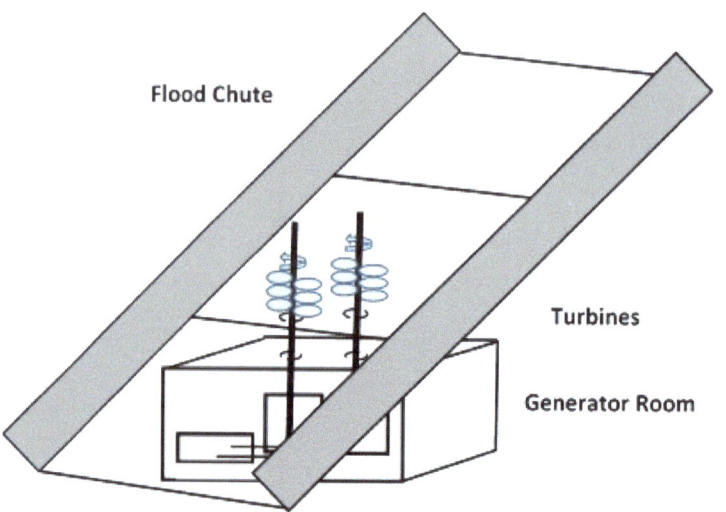

A.4.5
(Same as A.3.8, duplicated for this Appendix)

Generator and Power Line System
Two Turbines, Two Generators, 3 Phases x 2
Top View, Looking Down from Above

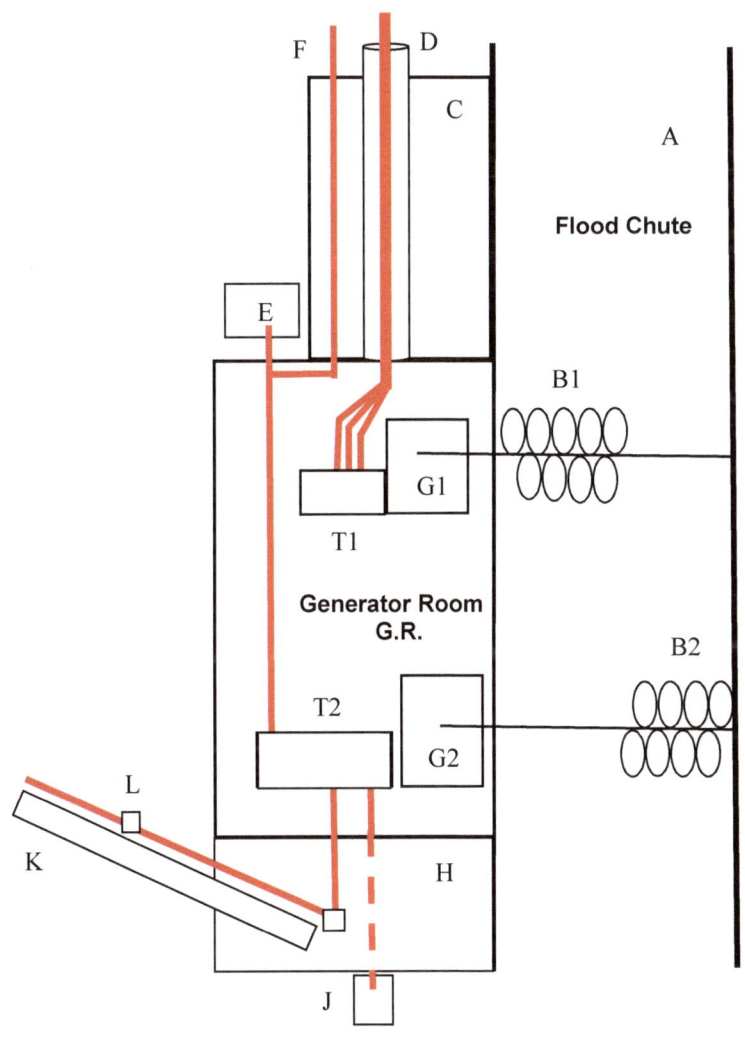

For meanings of each letter in this Diagram, see next page.

65

The Label Meanings for Drawing A.4.5 / A.3.8

A = <u>Flood Chute</u>
B1 = Turbine 1   B2 = Turbine 2

G.R. = <u>Generator Room</u>
G1 = Generator 1   G2 = Generator 2
T1 = Transformer 1   T1 = Transformer 2

C = <u>Maintenance Tunnel</u>
D = Conduit on roof, with Main Power Line
E = Elevator
F = Power Line for lights and doors in Maintenance Tunnel

H = <u>Shelf Extension / Top of Storage Room</u>
J = Primary Pump (at Absolute Ground, in front of Room)
K = Archimedes Screws for Water Conveyance
L = Gear Box for each Archimedes Screw

---

(1) All 3 phases of Power from Generator 1 will go up conduit, providing power for the public use.

(2) Generator 2 provides power for internal operations:
    P1 = for Primary Pump and Deeper Operations
    P2 = for Lights, Doors, Elevator
    P3 = for Archimedes Conveyances and Other Pumps

Appendix 5:
Filtration and Pumping Systems Drawings

Introduction

We can create clean drinking water, and deliver this clean water to the public, through a system of filters and pumps.

These filters and pumps can be placed anywhere, however the preferred placement is at the End Point. After all the flood waters are dropped into the Storage Container, the water will be sent to filters to be cleaned into drinking water. Then the water is sent back up to the surface using a series of pumps.

Filter Systems

There are numerous filter systems available. The filter method shown is a gravity filtration system, where the water falls through a series of filters. By the time the water reaches the bottom, the water will be clean enough for drinking.

The filter system is built adjacent to the storage container, at a much lower depth. Water flows from an opening at the bottom side wall into the filter system.

The filtration unit is a container, with a series of filters. Each filter is of successively finer filtration. Numerous types of filters exist and can be fitted into the system. There is no limit to the number of filters or materials used as filters in this system.

As a final purification mechanism, UV disinfection can be used. The UV disinfection can be placed along the traveling pipe which connects the filtered water to the primary pump. An additional UV disinfection system can be placed in the long-term storage of clean water at the surface.

A.5.1
Water Filtration System

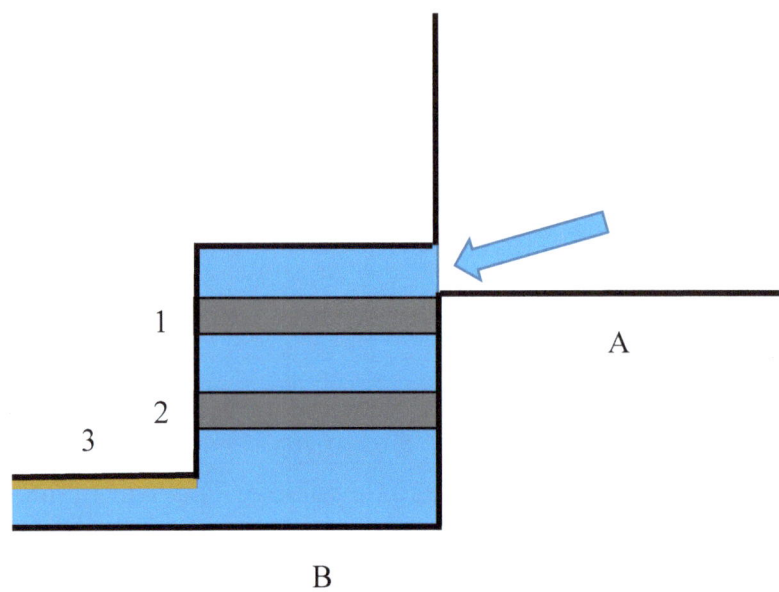

A = Flood Storage Container; showing bottom corner only

B = Filtration System
    1 = Filter #1: filtering size range of 1.0 to 5.0 microns
    2 = Filter #2: filtering size range of .25 to 1.0 microns
    3 = Filter #3: UV disinfection channel

Pump Systems

After the water has been filtered into clean drinking water, the water can be pumped to the surface for the public.

In this design, there are multiple pumps used in succession. The first is the Primary Pump, which brings the clean water from Absolute Bottom to the Main Level. Then at the Main Level, a series of secondary pumps will deliver the clean water to the surface.

Primary Pump

The Primary Pump is mechanical pump, which is powered by the same hydropower system at the End Point. Numerous mechanical pump designs are available, any of which are suitable for this system.

Archimedes Screw Conveyances

To get the clean water from the Main Level to the surface, we use a series of Archimedes Screws. These devices are commonly used for moving water, and will be effective for bringing water from the Main Level to the surface, and into a long-term storage container.

The Archimedes Screw Conveyances will also be powered by the hydropower system, with a power line and series of gear boxes lining the outside of the Archimedes Screw Tunnels.

A.5.2
Filtration System with Primary Pump

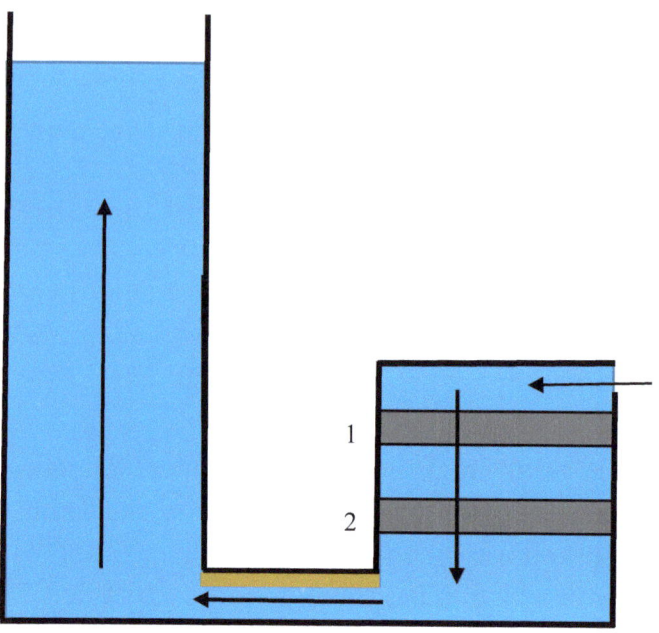

B = Filtration System
    1 = Filter #1: filtering size range of 1.0 to 5.0 microns
    2 = Filter #2: filtering size range of .25 to 1.0 microns
    3 = Filter #3: UV disinfection channel
C = Primary Pump

A.5.3
Archimedes Screw Conveyances
From Main Level, to the Surface

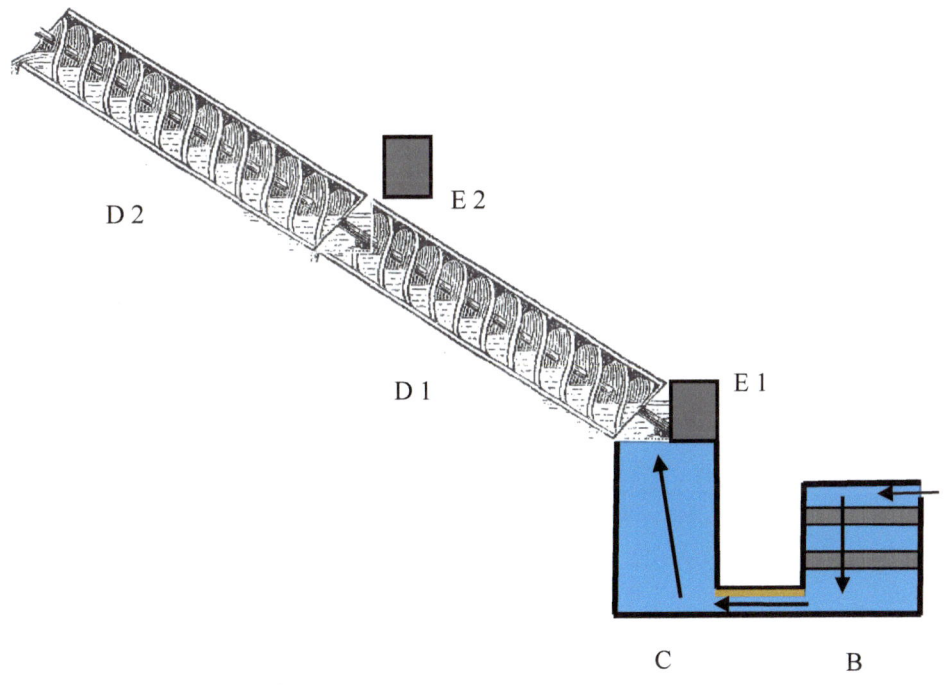

B = Filtration System
C = Primary Pump
D = Archimedes Screw Conveyance
E = Gear Box to Drive Archimedes Screw

A.5.4
Final Conveyance to Surface and Long-Term Storage

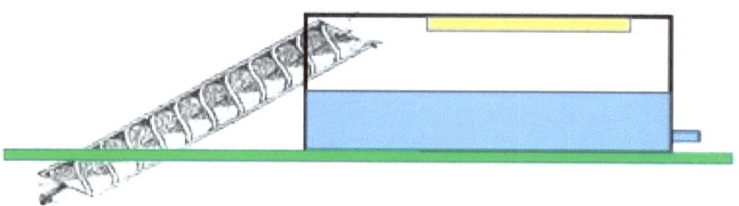

A.5.4 Final Conveyance to Surface and Long-Term Storage

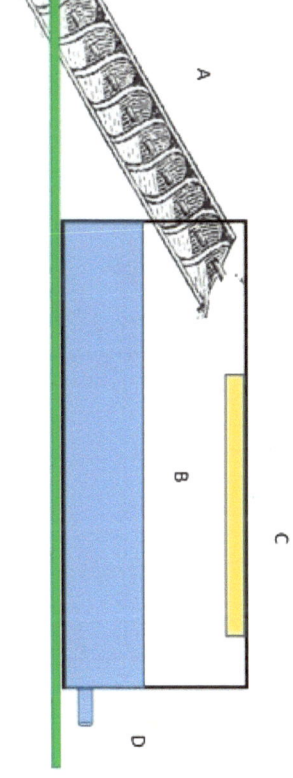

A = Final Archimedes Screw Conveyance
B = Long-Term Storage of Drinking Water
C = UV Disinfecting Bars
D = Outlet for Accesing Clean Water

A.5.5
Long-Term Storage Option B:
Below Ground

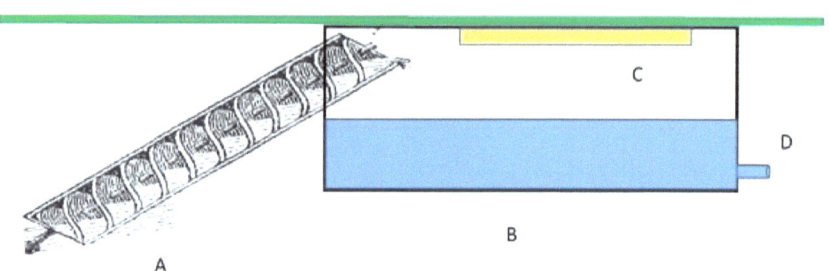

A = Final Archimedes Screw Conveyance

B = Long-Term Storage of Drinking Water

C = UV Disinfecting Bars

D = Outlet for Accesing Clean Water

A.5.6
Entire Filtration, Pumping System, and Long-Term Storage Altogether

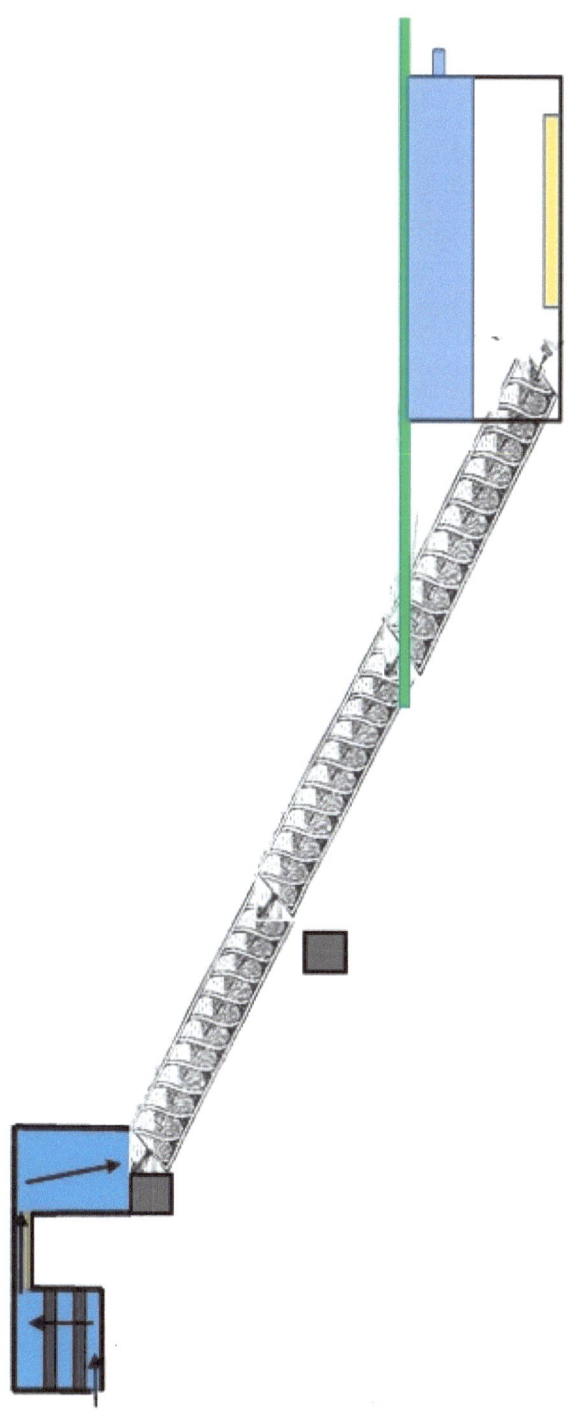

A.5.7
Power Lines for Pumping Systems.

Schematic Drawing
Looking Down from Above

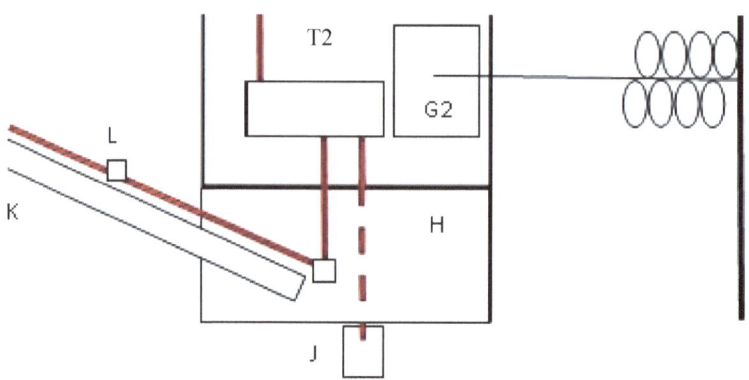

H = <u>Shelf Extension / Top of Storage Room</u>

G2 = Generator 2

T2 = Transformer Bank 2

J = Primary Pump (at Absolute Ground, in front of Room)

K = Archimdes Screws for Water Conveyance

L = Gear Box for each Archimedes Screw

Appendix 6:
Maintenance and Related Items
for Flood Control System

Introduction

In order to allow maintenance to be as easy as possible, every component, structure, and location in the flood control system will have easy access for maintenance. These designs are completely integrated into the entire flood control system.

Maintenance for Each Equipment

All equipment and structures in this flood control system will need to be maintained. Although this is an advanced system, some maintenance is required. Filters must be replaced, debris must be removed, cracks must be filled, and so forth. Therefore designs for maintenance are just as important as designs for flood control and clean water.

Maintenance designs are completely integrated into all aspects of this flood control system. Every component, every structure, will have easy access for inspection and maintenance.

The specific access points are summarized below. This is accompanied by illustrations.

Drawings of Maintenance Designs: Overview

Most of the Drawings in this Appendix are also placed in other Appendix sets. Yet it is important to reprint these same drawings here.

In this Appendix we are focusing on the Maintenance Access and related items. Therefore, review the writings regarding maintenance on these pages, then refer to the drawings. Go back and forth if needed.

Flood Chutes, Drainage Rings, and Maintenance Tunnels

1. Flood Chutes and their Maintenance Tunnels

Flood Chutes will be accessed from Maintenance Tunnels, which are built adjacent to the Chutes. Each Flood Chute in the system will have an accompanying Maintenance Tunnel, which is built adjacent to, and runs the entire length of, the Flood Chute.

Maintenance Doors are then installed at regular intervals. Workers can therefore access the Flood Chute at any location through the nearest door inside the Maintenance Tunnel.

2. Drainage Rings

Drainage Rings are also accessed in the Maintenance Tunnels, but from a different entrance. Specifically, each Drainage Ring is accessed through the Ring-Chute Connector.

Each Drainage Ring intersects the Maintenance Tunnel, before poking through the wall of the Flood Chute. The specific connection point is the Ring-Chute Connector, which has parallel horizontal holes for the pipe to lay inside.

In addition, the Ring-Chute Connector also has a top entrance. There is a simple man-hole with cover at the top of the Connector. The worker simply opens the cover, enters the Connector, and turns to the Drainage Ring Pipe.

Furthermore, due to the spoke and wheel design, there are several maintenance tunnels which cross each drainage ring. The worker can therefore enter any Ring Section from the Maintenance Tunnel on either side of that section.

A.6.1
Maintenance Tunnel, Flood Chute, Drainage Ring
(Same as A.2.11)

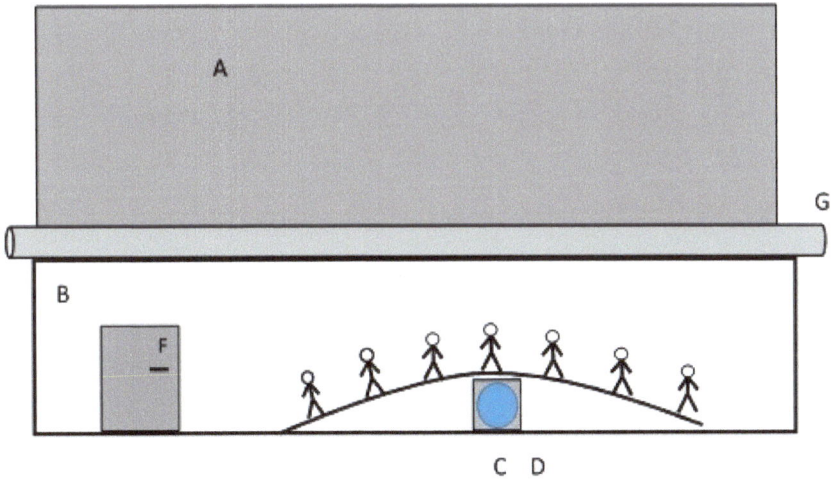

A = Flood Chute

B = Maintenance Tunnel

C = Drainage Ring

D = Ring-Chute Connector

E = Ramp over Ring-Chute Connector

F = Maintenance Door to Flood Chute

G = Main Power Line

End Point and Main Level
Including Turbines and Generators

3. <u>End Points</u>

The End Points of the System are the locations where the Flood Storage Containers are installed. Also at these locations will be: turbines, generators, filters, and pumps. The entire End Point location is buried deep underground.

To reach the general location of the End Point, workers use a simple freight elevator. This elevator will bring workers and equipment to the Main Level of the End Point.

4. <u>Turbines</u>

The Turbines are accessed through the Maintenance Doors of the Maintenance Tunnel near the Turbine. In the Maintenance Tunnel, doors are placed regularly which allow access to the Flood Chute. The final such door will be approximately 500 feet before the Turbines.

5. <u>Generator Room</u>

The Generator Room is a separate room built on top of the Main Level. Within this room are the generators, the transformers, and the initial power lines. The Generator Room is large room, with a door on the outside. Therefore, workers easily enter the Generator Room by first reaching the Main Level (using the elevator) then the entering the Door.

A.6.2
Main Level of End Point, Viewed from Above
(Same as A.3.6)

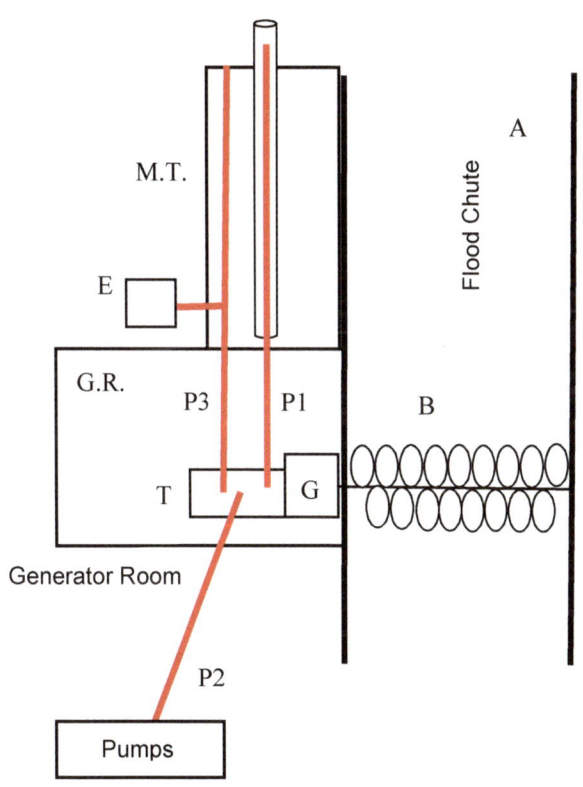

A = Flood Chute   B = Turbine   G = Generator
T = Transformer Bank   E = Elevator

P1 = Phase 1 of Power Line (Main)
P2 = Phase 2 of Power Line (Pumps)
P3 = Phase 3 of Power Line (Elevator, Lights, Doors)

G.R. = Generator Room   M.T. = Maintenance Tunnel

Storage Room, Filters, and Primary Pump Maintenance

6. Absolute Bottom and Pulley System

The End Points have three basic levels: Ceiling; Main Level; and Absolute Bottom. To reach the Main Level from the surface, workers use a simple freight elevator. Then, to go from the Main Level to Absolute Bottom, workers use a Pulley System. This Pulley System is the same device used for painters and repairmen outside large buildings. It is simple yet sturdy, and effective.

7. Filter Access

The filter structure will likely be the most often visited structure for the maintenance workers. The men will access the filters first through the pulley system, then through numerous doors.

Men will take pulley system from the Main Level down to the top of the Filter. At this point, there are numerous doors and attached ladders, for the workers to access any filter easily.

8. Primary Pump

The Primary Pump can be accessed from Absolute Bottom. Workers take the pulley system from the Main Level to Absolute Bottom. From there, an entrance door allows men to enter the pump system.

In addition, we can also install attached ladders (either outside or inside the pump structure).

A.6.3 (Same as A.3.10)
Pulley Systems and Access Doors to Reach Filters, Pumps, Storage.

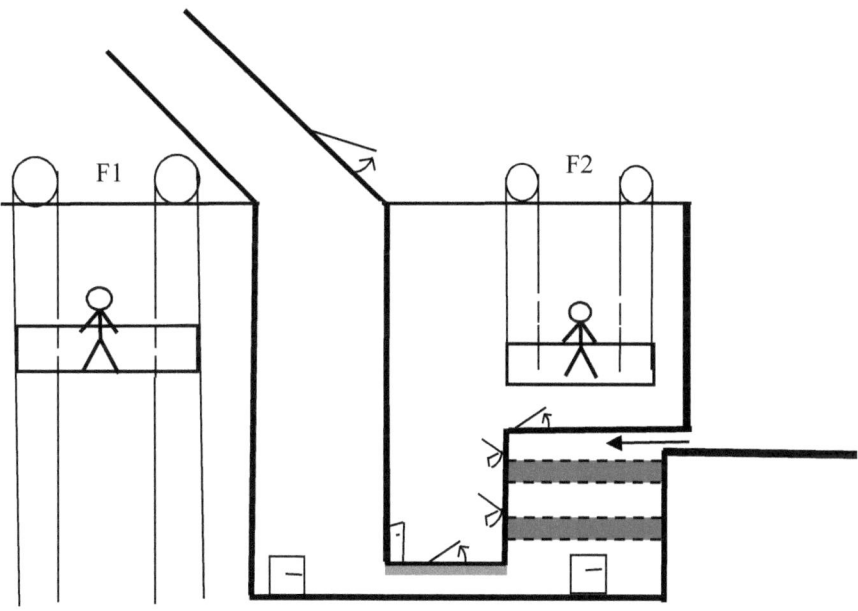

A = Flood Storage Container
B = Maintenance Storage Room
C = Filter System*
D = Primary Pump*
E = Archimedes Screw Conveyance

F1 = Pulley System to Absolute Ground*
F2 = Pulley System to Top of Filter*

G = Main Door to Pump Structure
H = Main Door to Filter Structure
J = Doors to Filters and Filter System
K = Side Door to Primary Pump
L = Maintenance Door to Archimedes Screw

*Remember that Pulley Systems are on the *outside* of the Maintenance Storage Room. Also remember that the Filter and the Primary Pump will sit in front of the Storage Room.

Maintenance for Archimedes Screw Conveyance

9. Archimedes Screw Conveyance: Motors, Gears, Power Lines

For the Archimedes Screw Conveyance Systems, most of the maintenance will be outside the conveyance structure. It is the gears, motors, and power lines which are more likely to need repair than the Archimedes Screw itself.

The gear boxes are placed outside the Conveyance Structures, therefore accessing the motors and gears will be simple. Similarly, the Power Lines are laid along the outside of the Conveyance Structures. Therefore, the power lines can be accessed and replaced easily.

10. Archimedes Screw Conveyance: The Blades of the Screw

The Archimedes Screw itself will not require much maintenance. However, when needed each Archimedes Screw Conveyance can be accessed through side doors. In this case, the men will not need to enter the structure, they can simply open up the side panels to inspect and do the repair they need.

The first and last conveyance will be the easiest to access. The first conveyance begins within the End Point. Therefore the men can open up large side doors (up to 5 feet long) to inspect and repair the first conveyance.

The last conveyance will end above ground, and tall enough to pour into a long-term storage container. Therefore, again, it is easy for men to walk the meadow, open a side door, and perform maintenance.

For the conveyance structures within the earth, there must be space for the worker to move up the side of the conveyance, and therefore crawl spaces can be built along the conveyance from the main level to the surface. (Similar to maintenance tunnels along the flood chutes, yet smaller and simpler).

A.6.4
Archimedes Screw Conveyance, Gear Box, and Power Line Easily Accessed on Main Level of End Point

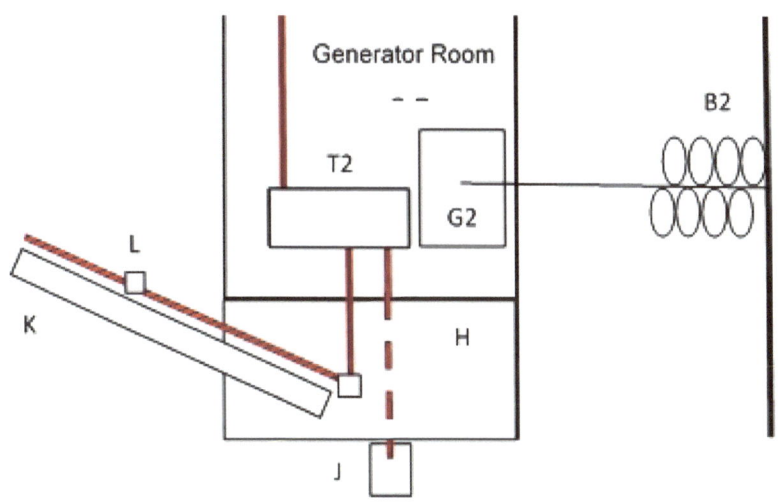

H = <u>Shelf Extension / Top of Storage Room</u>

G2 = Generator 2

T2 = Transformer Bank 2

J = Primary Pump (at Absolute Ground, in front of Room)

K = Archimedes Screws for Water Conveyance

L = Gear Box for each Archimedes Screw

A.6.5
(same as A.5.4)
Final Conveyance to Surface.

Easy Access Above Ground for Maintenance of Conveyance Screw and Storage

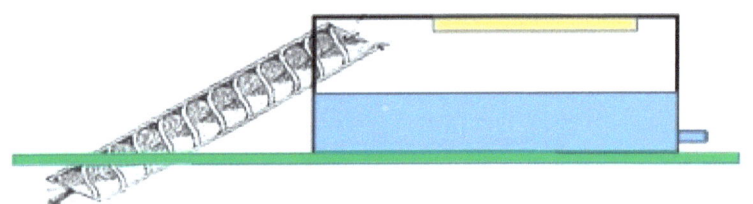

Maintenance Access for Storage Room

11. <u>Storage Room</u>

The Storage Room is a very large room, which is adjacent to the Flood Storage Container. The top of this Room is an Extension of the Main Level, on which the Archimedes Screw is placed, as well as the Pulley System which reaches Absolute Bottom.

The Storage Room is accessed by a large door at Absolute Bottom. Workers use the Pulley System to go from Main Level to Absolute Bottom, and back up again. The large doors (wide and tall) allow equipment of any size to be stored in the Room, and removed again when needed.

The equipment, of course, can be brought up to the Main Level using the same Pulley System.

A.6.6
Pulley Platform, and Access Doors to Storage Room
(Same as A.3.9)

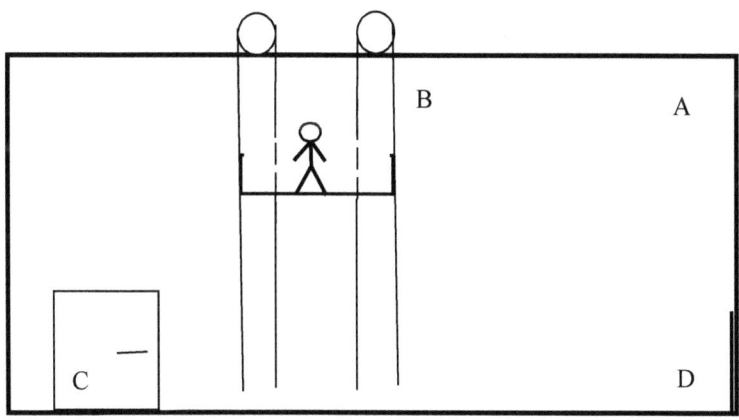

- A = Maintenance Storage Room
- B = Pulley System for Delivering Men and Equipment (from Main Level to Absolute Ground and back up)
- C = Door to Maintenance Storage Room
- D = Door between Maintenance Room and Flood Storage Container (accessed inside Storage Room)

Maintenance Access for Flood Container

12. <u>Flood Storage Container: Main Access</u>

The Flood Storage Container has one access door: from the Maintenance Storage Room. The Flood Storage Container may also be accessed via attached ladders inside the container.

Inside the Storage Room, there is a very large door which connects the Storage Room to the Flood Storage Container. This door can be mechanically opened as overflow when the Flood Waters exceed the capacity of the Flood Container. Yet these same doors will allow workers to access the empty Container, and bring any equipment they need.

Therefore, when a worker wants to inspect and repair the (now empty) Flood Storage Container, he will first go to Absolute Bottom, then go into the Storage Room, and then finally into the Flood Storage Container.

He will bring all equipment he needs through those doors.

Note that this door will appear to be at different heights for each of these two rooms. This is because the bottom of the Flood Storage Container is higher than the bottom of the Maintenance Storage Room. (Remember the gravity filter system is lower than Flood Container, and both filter and storage room sit on absolute bottom).

Therefore a ladder or ramp must be installed on one side or the other, to access this door. For example, a ramp leads upward from the Storage Room floor, up to the Door. When the workers open the door, they will walk into the Flood Container, onto the floor of that Container.

A.6.7
Access Door between
Maintenance Storage Room and Flood Container.

Viewing: Standing at Absolute Bottom, looking at "front" of End Point.

(This is also prior to Pumps and Filters being installed).

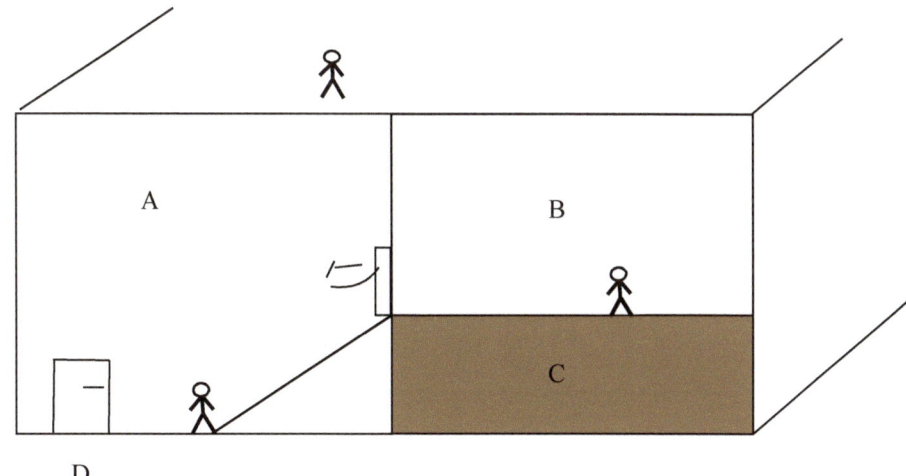

A = Maintenance Storage Room
B = Flood Storage Container
C = Dirt and Rocks to Support Flood Container
D = Door to Storage Room
E = Ramp inside, leading to Flood Container Door
F = Access Door to Flood Container

13. Flood Storage Container: Attached Ladders
An alternate access can be attached ladders inside the Flood Storage Container. Workers on the Main Level walk to the edge of the Flood Storage Container, then down the ladder into the container.

A.6.8
Left Wall of Empty Flood Container, Viewed from the Inside.

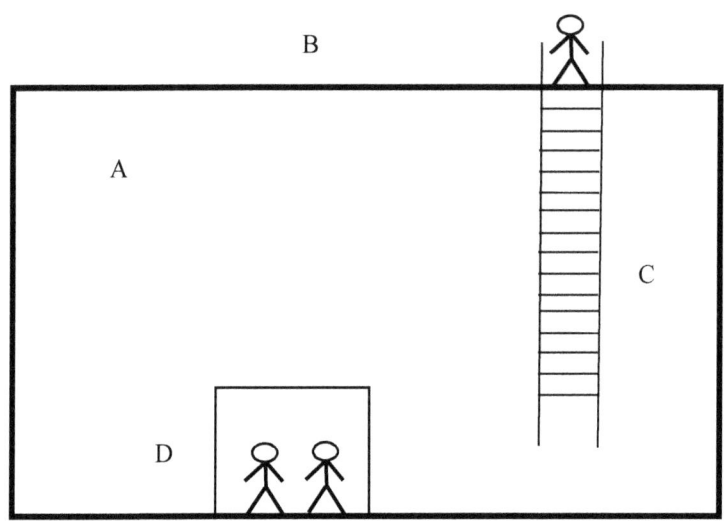

A = Left Inside Wall of Flood Container

B = Main Level; Surface Extension; top of Storage Room

C = Attached Ladder, for workers to go from Main Level
 to Flood Container Floor

D = Large Door between Flood Container and Storage Room

Appendix 7
Structural Boundaries and Support of the End Point

Introduction

The entire "End Point" will be contained in a type of concrete box. This "Structural Boundary" is essentially a very large room, in which all End Point Systems will be contained.

Within the Structural Boundary the workers can move about, and all the system structures will be protected. Outside the Structural Boundary will be the earth.

Walls and Ceiling of the Structural Boundary

The Structural Boundary is essentially a very large room, with walls of varying height, and a ceiling that is level.

Most of the walls will simply be extensions of the structures already part of the End Point. Other walls will be built specifically to be the structural boundaries.

The "ceiling" to the structural boundaries is very important. Remember that the workers will want to walk on the main level, therefore a supported ceiling must exist. The ceiling should be taller than the generator room.

The front wall is also important. This may be the only wall which extends fully from Absolute Ground to the Ceiling of the Structural Boundary. Also, this wall must extend several feet beyond all structures. Remember that we want space at absolute ground for workers to walk in front of the filters and primary pump.

The Structural Boundary is essentially a room, a very large box, with supporting walls and pillars. However, there will also be a few essential holes. These openings must exist for: Flood Chute; Maintenance Tunnel; Archimedes Conveyance; and Elevator.

In total, this "Structural Boundary" is the concrete "room" in which all of the End Point Systems are contained. It will protect everything inside, and allow the men to move about for maintenance.

The following drawings will help to illustrate the design.

A.7.1
Structural Boundaries of End Point, Outer View of the Entire Containment

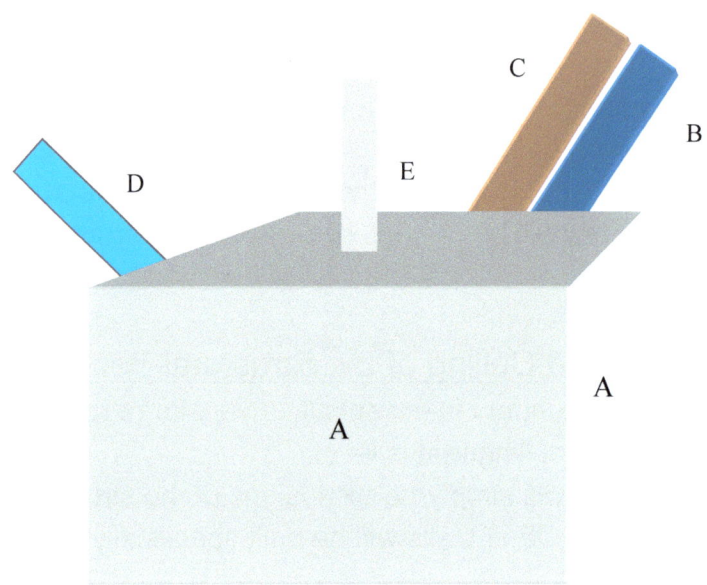

A = Structural Boundary for End Point Systems
B = Flood Chute
C = Maintenance Tunnel
D = Archimedes Conveyance
E = Elevator

A.7.2
(Omitted in Abridged Version)

A.7.3
Right Side View of Structural Boundaries, with outer right wall "removed"

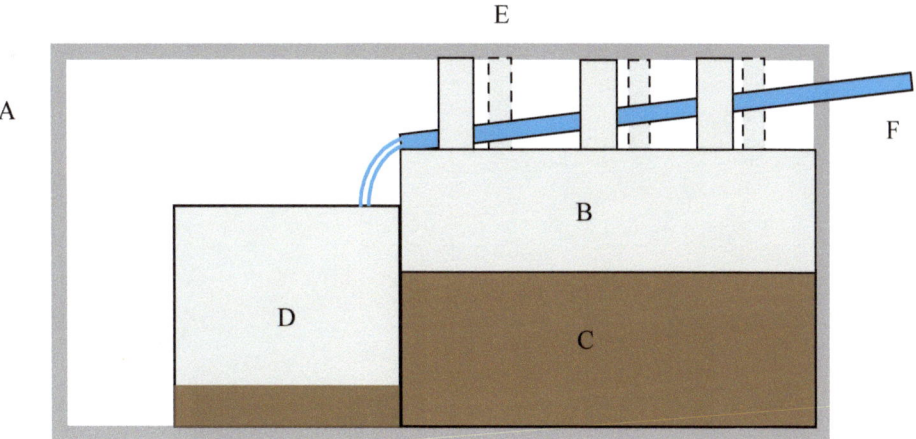

A = Structure Boundary Containment for End Point Systems

B = Main Level: Flat Concrete Support for Generator, Pillars, etc.

C = Dirt and Rocks, for Support of Main Level

D = Flood Storage Container

E = Supporting Posts: to Support Ceiling Above Main Level

F = Flood Chute; Enters through Hole in Back Wall.

A.7.4
(Omitted in Abridged Version)

A.7.5
End Point Structural Boundaries, #2
Front Side View of Surface Extension

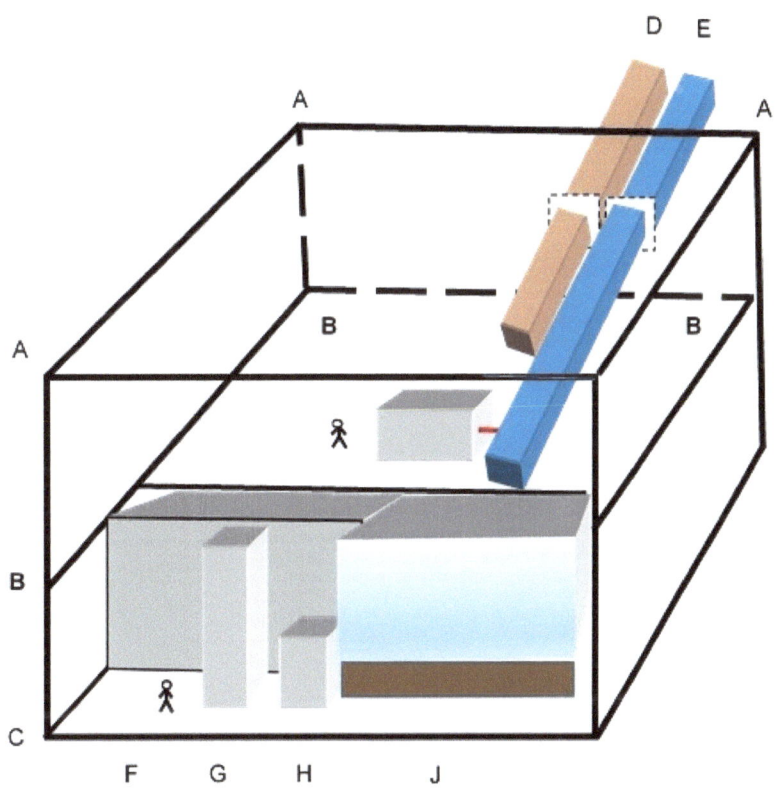

A = End Point Ceiling   B = Main Level
       C = Absolute Bottom

D = Maintenance Tunnel   E = Flood Chute

F = Storage Room   G = Pump
H = Filter System   J = Flood Container

This Page
Intentionally Left Blank

About the Designer

My name is Mark Fennell. I am a scientist and inventor. I have modeled myself after Ben Franklin, Thomas Jefferson, Leonardo Da Vinci, and Nikola Tesla.

A few accomplishments worth noting:
- Authored and published over 60 books
- Created over 30 inventions
- Discovered solutions which greatest scientists could not (including Einstein, Feynman, Hawking)

Most of my scientific discoveries and inventions are related to:
- Electrical Power
- Energy Science
- Gravity
- Motion and Friction
- Unified Energy and Quantum Science

You can learn more about my discoveries and books at:
https://www.amazon.com/author/markfennellvisionary

You can read more about me and my work at:
http://markfennellvisionary.com

Regarding the design of this Flood Control System, know that I have studied civil engineering for years. This includes water management systems, tunnels, mining, and electrical power.

Also note that I have personally experienced flooding in Houston, and Dallas; throughout much of Ohio; and several areas of Indiana. I know the geography, the topography, and the existing inadequate drainage systems in each of those areas.

This design will work, and will prevent any future flooding in metropolitan areas, when installed and maintained properly.

The only reason I provide this "About Inventor" page is so that city planners and engineers will look at this proposal with respect. This page is *not* written to talk about me, it is written so that the regional planners will seriously consider building this Design.

Use this design. Modify it. Just…build it.

Mark Fennell
markpoet@aol.com

www.ingramcontent.com/pod-product-compliance
Lightning Source LLC
Chambersburg PA
CBHW051155220526
45473CB00003B/781